Farm Engines
and How to Run Them

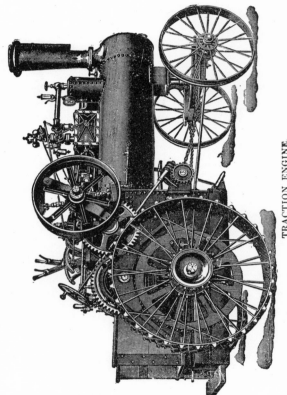

TRACTION ENGINE.

Farm Engines
and How to Run Them

A Simple, Practical Handbook for
Experts and Amateurs

James H. Stephenson

THE LYONS PRESS
Guilford, Connecticut
An imprint of The Globe Pequot Press

The Lyons Press is an imprint of The Globe Pequot Press.

10 9 8 7 6 5 4 3 2 1

Printed in the United States of America

Originally published in 1903 by Frederick J. Drake & Co.

Library of Congress Cataloging-in-Publication Data

Farm engines and how to run them / James H. Stephenson.
        p. cm.
Includes index.
ISBN 1-59228-303-9 (trade paper)
1. Farm engines—Juvenile literature.
Tj712.F37 2004
631.3'72—dc22

                                                2004046463

PREFACE

This book makes no pretensions to originality. It has taken the best from every source. The author believes the matter has been arranged in a more simple and effective manner, and that more information has been crowded into these pages than will be found within the pages of any similar book.

The professional engineer, in writing a book for young engineers, is likely to forget that the novice is unfamiliar with many terms which are like daily bread to him. The present writers have tried to avoid that pitfall, and to define each term as it naturally needs definition. Moreover, the description of parts and the definitions of terms have preceded any suggestions on operation, the authors believing that the young engineer should become thoroughly familiar with his engine and its manner of working, before he is told what is best to do and not to do. If he is forced on too fast he is likely to get mixed. The test questions at the end of Chapter III. will show how perfectly the preceding pages have been mastered, and the student is not ready to go on till he can answer all these questions readily.

The system of questions and answers has its uses and its limitations. The authors have tried to use that system where it would do most good, and employ the straight narrative discussion method where questions could not help and would only interrupt the progress of thought. Little technical matter has been introduced, and that only for practical purposes. The authors have had traction engines in mind for the most part, but the directions will apply equally well to any kind of steam engine.

The thanks of the publishers are due to the various traction engine and threshing machine manufacturers for cuts and information, and especially to the *Thresher-men's Review* for ideas contained in its "Farm Engine Economy," to the J. I. Case Threshing Machine Co. for the use of copyrighted matter in their "The Science of Successful Threshing," and to the manager of the Columbus Machine Co. for valuable personal information furnished the authors on gasoline engines and how to run them. The proof has been read and corrected by Mr. T. R. Butman, known in Chicago for 25 years as one of the leading experts on engines and boilers, especially boilers.

THE
YOUNG ENGINEERS' GUIDE

CHAPTER I.

BUYING AN ENGINE.

There are a great many makes of good engines on the market to-day, and the competition is so keen that no engine maker can afford to turn out a very poor engine. This is especially true of traction engines. The different styles and types all have their advantages, and are good in their way. For all that, one good engine may be valueless for you, and there are many ways in which you may make a great mistake in purchasing an engine. The following points will help you to choose wisely:

1. Consider what you want an engine for. If it is a stationary engine, consider the work to be done, the space it is to occupy, and what conveniences will save your time. Remember, TIME IS MONEY, and that means that SPACE IS ALSO MONEY. Choose the kind of engine that will be most convenient for the position in which you wish to place it and the purpose or purposes for which you wish to use it. If buying a traction engine, consider also the roads and an engine's pulling qualities.

2. If you are buying a traction engine for threshing, the first thing to consider is FUEL. Which will be cheapest for you, wood, coal or straw? Is economy of fuel much of an object with you—one that will justify you in greater care and more scientific study of your engine? Other things being equal, the direct flue, firebox,

locomotive boiler and simple engine will be the best, since they are the easiest to operate. They are not the most economical under favorable conditions, but a return flue boiler and a compound engine will cost you far more than the possible saving of fuel unless you manage them in a scientific way. Indeed, if not rightly managed they will waste more fuel than the direct flue locomotive boiler and the simple engine.

3. Do not try to economize on the size of your boiler, and at the same time never get too large an engine. If a 6-horse power boiler will just do your work, an 8-horse power will do it better and more economically, because you won't be overworking it all the time. Engines should seldom be crowded. At the same time you never know when you may want a higher capacity than you have, or how much you may lose by not having it. Of course you don't want an engine and boiler that are too big, but you should always allow a fair margin above your anticipated requirements.

4. Do not try to economize on appliances. You should have a good pump, a good injector, a good heater, an extra steam gauge, an extra fusible plug ready to put in, a flue expander and a beader. You should also certainly have a good force pump and hose to clean the boiler, and the best oil and grease you can get. Never believe the man who tells you that something not quite the best is just as good. You will find it the most expensive thing you ever tried—if you have wit enough to find out how expensive it is.

5. If you want my personal advice on the proper engine to select for various purposes, I should say by all means get a gasoline engine for small powers about the farm, such as pumping, etc. It is the quickest to start, by far the most economical to operate, and the simplest to manage. The day of the small steam engine is past and will never return, and ten gasoline engines of this kind are sold for every steam engine put out. If you want a traction engine for threshing, etc., stick to steam. Gasoline engines are not very good hill climbers because the application of power is not steady enough; they are

not very good to get out of mud holes with for the same reason, and as yet they are not perfected for such purposes. You might use a portable gasoline engine, however, though the application of power is not as steady as with steam and the flywheels are heavy. In choosing a traction steam engine, the direct flue locomotive boiler and simple engine, though theoretically not so economical as the return flue boiler and compound engine, will in many cases prove so practically because they are so much simpler and there is not the chance to go wrong with them that there is with the others. If for any reason you want a very quick steamer, buy an upright. If economy of fuel is very important and you are prepared to make the necessary effort to secure it, a return flue boiler will be a good investment, and a really good compound engine may be. Where a large plant is to be operated and a high power constant and steady energy is demanded, stick to steam, since the gasoline engines of the larger size have not proved so successful, and are certainly by no means so steady; and in such a case the exhaust steam can be used for heating and for various other purposes that will work the greatest economy. For such a plant choose a horizontal tubular boiler, set in masonry, and a compound engine (the latter if you have a scientific engineer).

In general, in the traction engine, look to the convenience of arrangement of the throttle, reverse lever, steering wheel, friction clutch, independent pump and injector, all of which should be within easy reach of the footboard, as such an arrangement will save annoyance and often damage when quick action is required.

The boiler should be well set; the firebox large, with large grate surface if a locomotive type of boiler is used, and the number of flues should be sufficient to allow good combustion without forced draft. A return flue boiler should have a large main flue, material of the required 5-16-inch thickness, a mud drum, and four to six handholes suitably situated for cleaning the boiler. There should be a rather high average boiler pressure, as high pressure is more economical than low. For a simple engine, 80 pounds and for a compound 125 pounds should be minimum.

A stationary engine should have a solid foundation built by a mason who understands the business, and should be in a light, dry room—never in a dark cellar or a damp place.

Every farm traction engine should have a friction clutch.

CHAPTER II.

The first boilers were made as a single cylinder of wrought iron set in brick work, with provision for a fire under one end. This was used for many years, but it produced steam very slowly and with great waste of fuel.

The first improvement to be made in this was a fire flue running the whole length of the interior of the boiler, with the fire in one end of the flue. This fire flue was entirely surrounded by water.

Then a boiler was made with two flues that came together at the smoke-box end. First one flue was fired and then the other, alternately, the clear heat of one burning the smoke of the other when it came into the common passage.

The next step was to introduce conical tubes by which the water could circulate through the main fire flue (Galloway boiler).

FIG. 1. ORR & SEMBOWER'S STANDARD HORIZONTAL BOILER, WITH FULL-ARCH FRONT SETTING.

The object of all these improvements was to get larger heating surface. To make steam rapidly and economically, the heating surface must be as large as possible.

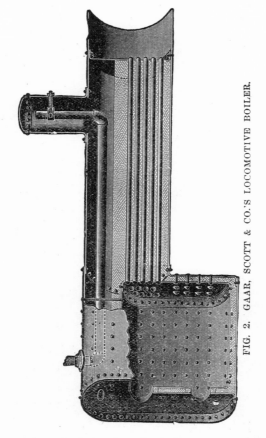

FIG. 2. GAAR, SCOTT & CO.'S LOCOMOTIVE BOILER.

But there is a limit in that the boiler must not be cumbersome, it must carry enough water, and have sufficient space for steam.

The stationary boiler now most commonly used is cylindrical, the fire is built in a brick furnace under the sheet and returns through fire tubes running the length of the boiler. (Fig. 1.)

LOCOMOTIVE FIRE TUBE TYPE OF BOILER.

The earliest of the modern steam boilers to come into use was the locomotive fire tube type, with a special firebox. By reference to the illustration (Fig. 2) you will see that the boiler cylinder is perforated with a number of tubes from 2 to 4 inches in diameter running from the large firebox on the left, through the boiler cylinder filled

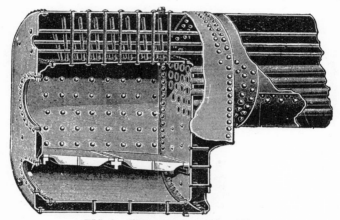

FIG. 3. THE HUBER FIRE BOX.

with water, to the smoke-box on the right, above which the smokestack rises.

It will be noticed that the walls of the firebox are double, and that the water circulates freely all about the firebox as well as all about the fire tubes. The inner walls of the firebox are held firmly in position by stay bolts, as will be seen in Fig. 3, which also shows the position of the grate.

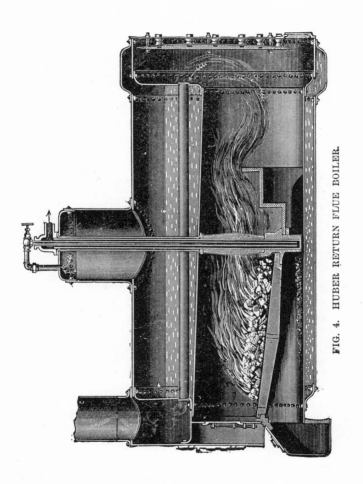

FIG. 4. HUBER RETURN FLUE BOILER.

RETURN FLUE TYPE OF BOILER.

The return flue type of boiler consists of a large central fire flue running through the boiler cylinder to the smoke box at the front end, which is entirely closed. The smoke passes back through a number of small tubes, and the smokestack is directly over the fire at the rear of the boiler, though there is no communication between the fire at the rear of the boiler and it except through the main flue to the front and back through the small return flues. Fig. 4 illustrates this type of boiler, though it shows but one return flue. The actual number may be seen by the sectional view in Fig. 5.

The fire is built in one end of the main flue, and is entirely surrounded by water, as will be seen in the illustration. The long passage for the flame and heated gases enables the water to absorb a maximum amount of the heat of combustion. There is also an element of safety in

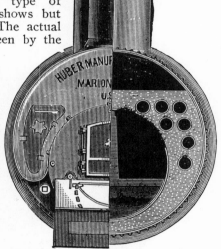

FIG. 5. SECTION VIEW OF HUBER RETURN FLUE BOILER.

this boiler in that the small flues will be exposed first should the water become low, and less damage will be done than if the large crown sheet of the firebox boiler is exposed, and this large crown sheet is the first thing to be exposed in that type of boiler.

WATER TUBE TYPE OF BOILER.

The special difference between the fire tube boiler and the water tube boiler is that in the former the fire passes through the tubes, while in the latter the water is in the tubes and the fire passes around them.

In this type of boiler there is an upper cylinder (or

FIG. 6. FREEMAN VERTICAL BOILER.

more than one) filled with water; a series of small tubes running at an angle from the front or fire door end of the upper cylinder to a point below and back of the grates, where they meet in another cylinder or pipe, which is connected with the other end of the upper cylinder. The portions of the tubes directly over the fire will be hottest, and the water here will become heated and. rise to the front end of the upper cylinder, while to fill the space left, colder water is drawn in from the back pipe, from the rear end of the upper cylinder, down to the lower ends of the water tubes, to pass along up through them to the front end again.

This type of boiler gives great heating surface, and since the tubes are small they will have ample strength with much thinner walls. Great freedom of circulation is important in this type of boiler, there being no contracted cells in the passage. This is not adapted for a portable engine.

UPRIGHT OR VERTICAL TYPE OF BOILER.

In the upright type of boiler the boiler cylinder is placed on end, the fire is built at the lower end, which is a firebox surrounded by a water jacket, and the smoke and gases of combustion rise straight up through vertical fire flues. The amount of water carried is relatively small, and the steam space is also small, while the heating surface is relatively large if the boiler is sufficiently tall. You can get up steam in this type of boiler quicker than in any other, and in case of the stationary engine, the space occupied is a minimum. The majority of small stationary engines have this type of boiler, and there is a traction engine with upright boiler which has been widely used, but it is open to the objection that the upper or steam ends of the tubes easily get overheated and so become leaky. There is also often trouble from mud and scale deposits in the water leg, the bottom area of which is very small.

DEFINITION OF TERMS USED IN CONNECTION WITH BOILERS.

Shell—The main cylindrical steel sheets which form the principal part of the boiler.

Boiler-heads—The ends of the boiler cylinder.

Tube Sheets—The sheets in which the fire tubes are inserted at each end of the boiler.

Fire-box—A nearly square space at one end of a boiler, in which the fire is placed. Properly it is surrounded on all sides by a double wall, the space between the two shells of these walls being filled with water. All flat surfaces are securely fastened by stay bolts and crown bars, but cylindrical surfaces are self-bracing.

Water-leg—The space at sides of fire-box and below it in which water passes.

Crown-sheet—The sheet of steel at the top of the fire-box, just under the water in the boiler. This crown sheet is exposed to severe heat, but so long as it is covered with water, the water will conduct the heat away, and the metal can never become any hotter than the water in the boiler. If, however, it is not covered with water, but only by steam, it quickly becomes overheated, since the steam does not conduct the heat away as the water does. It may become so hot it will soften and sag, but the great danger is that the thin layer of water near this overheated crown sheet will be suddenly turned into a great volume of steam and cause an explosion. If some of the pressure is taken off, this overheated water may suddenly burst into steam and cause an explosion, as the safety valve blows off, for example (since the safety valve relieves some of the pressure).

Smoke-box—The space at the end of the boiler opposite to that of the fire, in which the smoke may accumulate before passing up the stack in the locomotive type, or through the small flues in the return type of boiler.

Steam-dome—A drum or projection at the top of the boiler cylinder, forming the highest point which the steam can reach. The steam is taken from the boiler through piping leading from the top of this dome, since at this point it is least likely to be mixed with water, either through foaming or shaking up of the boiler. Even under normal conditions the steam at the top of the dome is drier than anywhere else.

Mud-drum—A cylindrical-shaped receptacle at the bottom of the boiler similar to the steam-dome at the top, but not so deep. Impurities in the water accumulate here, and it is of great value on a return flue boiler. In a locomotive boiler the mud accumulates in the water leg, below the firebox.

Man-holes—Are large openings into the interior of a boiler, through which a man may pass to clean out the inside.

Hand-holes—Are smaller holes at various points in the boiler into which the nozzle of a hose may be introduced for cleaning out the interior. All these openings must be securely covered with steam-tight plates, called man-hole and hand-hole plates.

A boiler jacket—A non-conducting covering of wood, plaster, hair, rags, felt, paper, asbestos or the like, which prevents the boiler shell from cooling too rapidly through radiation of heat from the steel. These materials are usually held in place against the boiler by sheet iron. An intervening air-space between the jacket and the boiler shell will add to the efficiency of the jacket.

A steam-jacket—A space around an engine cylinder or the like which may be filled with live steam so as to keep the interior from cooling rapidly.

Ash-pit—The space directly under the grates, where the ashes accumulate.

Dead-plates—Solid sheets of steel on which the fire lies the same as on the grates, but with no openings through to the ash-pit. Dead-plates are sometimes used to prevent cold air passing through the fire into the flues, and are common on straw-burning boilers. They should seldom if ever be used on coal or wood firing boilers.

Grate Surface—The whole space occupied by the grate-bars, usually measured in square feet.

Forced Draft—A draft produced by any means other than the natural tendency of the heated gases of combustion to rise. For example, a draft caused by letting steam escape into the stack.

Heating Surface—The entire surface of the boiler exposed to the heat of the fire, or the area of steel or iron sheeting or tubing, on one side of which is water and on the other heated air or gases.

Steam-space—The cubical contents of the space which may be occupied by steam above the water.

Water-space—The cubical contents of the space occupied by water below the steam.

Diaphragm-plate—A perforated plate used in the domes of locomotive boilers to prevent water dashing into the steam supply pipe. A dry-pipe is a pipe with small perforations, used for taking steam from the steam-space, instead of from a dome with diaphragm-plate.

THE ATTACHMENTS OF A BOILER.*

Before proceeding to a consideration of the care and management of a boiler, let us briefly indicate the chief working attachments of a boiler. Unless the nature and uses of these attachments are fully understood, it will be impossible to handle the boiler in a thoroughly safe and scientific fashion, though some engineers do handle boilers without knowing all about these attachments. Their ignorance in many cases costs them their lives and the lives of others.

The first duty of the engineer is to see that the boiler is filled with water. This he usually does by looking at the glass water-gauge.

THE WATER GAUGE AND COCKS.

There is a cock at each end of the glass tube. When these cocks are open the water will pass through the lower into the glass tube, while steam comes through the other. The level of the water in the gauge will then be the same as the level of the water in the boiler,

TWO-ROD WATER GAUGE.

and the water should never fall out of sight below the lower end of the glass, nor rise above the upper end.

*Unless otherwise indicated, cuts of fittings show those manufactured by the Lunkenheimer Co., Cincinnati, Ohio.

Below the lower gauge cock there is another cock used for draining the gauge and blowing it off when there is a pressure of steam on. By occasionally opening this cock, allowing the heated water or steam to blow through it, the engineer may always be sure that the passages into the water gauge are not stopped up by any means. By closing the upper cock and opening the lower, the passage into the lower may be cleared by blowing off the drain cock; by closing the lower gauge cock and opening the upper the passage from the steam space may be cleared and tested in the same way when the drain cock is opened. If the glass breaks, both upper and lower gauge cocks should be closed instantly.

GAUGE OR TRY COCK.

In addition to the glass water gauge, there are the try-cocks for ascertaining the level of the water in the boiler. There should be two to four of these. They open directly out of the boiler sheet, and by opening them in turn it is possible to tell approximately where the water stands. There should be one cock near the level of the crown sheet, or slightly above it, another about the level of the lower gauge cock, another about the middle of the gauge, another about the level of the upper gauge, and still another, perhaps, a little higher. But one above and one below the water line will be sufficient. If water stands above the level of the cock, it will blow off white mist when opened; if the cock opens from steam-space, it will blow off blue steam when opened.

The try-cocks should be opened from time to time in order to be sure the water stands at the proper level in the boiler, for various things may interfere with the working of the glass gauge. Try-cocks are often called gauge cocks.

TRY COCK.

THE STEAM GAUGE.

The steam gauge is a delicate instrument arranged so as to indicate by a pointer the pounds of pressure which the steam is exerting within the boiler. It is extremely important, and a defect in it may cause much damage.

The steam gauge was invented in 1849 by Eugene Bourdon, of France. He discovered that a flat tube bent in a simple curve, held fast at one end, would expand and contract if made of proper spring material, through the pressure of the water within the tube. The free end operates a clock-work that moves the pointer.

PRESSURE GAUGE.

It is important that the steam gauge be attached to the boiler by a siphon, or with a knot in the tube, so that

STEAM GAUGE SIPHON.

the steam may operate on water contained in the tube, and the water cannot become displaced by steam, since steam might interfere with the correct working of the gauge by expanding the gauge tube through its excessive heat.

Steam gauges frequently get out of order, and should be tested occasionally. This may conveniently be done by attaching them to a boiler which has a correct gauge already on it. If both register alike, it is probable that both are accurate.

There are also self-testing steam gauges. With all pressure off, the pointer will return to O. Then a series of weights are arranged which may be hung on the gauge and cause the pointer to indicate corresponding numbers. The chief source of variation is in the loosening of the indicator needle. This shows itself usually when the pressure is off and the pointer does not return exactly to zero.

FRONT CYLINDER COCK.

SAFETY VALVE.

The safety valve is a valve held in place by a weighted lever* or by a spiral spring (on traction engines) or some similar device, and is adjustable by a screw or the like so that it can be set to blow off at a given pressure of steam, usually the rated pressure of the boiler, which on traction en-

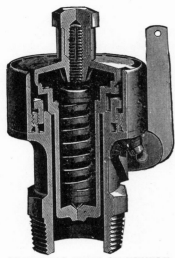

SECTIONAL VIEW OF KUNKLE POP VALVE.

SAFETY VALVE.

gines is from 110 to 130 pounds. The valve is supplied with a handle by which it can be opened, and it should be opened occasionally to make sure it is working all right. When it blows off the steam gauge should be noted to see that it agrees with the pressure for which the safety valve was set. If they do not agree, something is wrong; either the safety valve does not work freely, or the steam gauge does not register accurately.

The cut shows the Kunkle safety valve. To set it, unscrew the jam nut and apply the key to the pressure screw. For more pressure, screw down; for less, unscrew. After having the desired pressure, screw the jam

*This kind of safety valve is now being entirely discarded as much more dangerous than the spring or pop valve.

nut down tight on the pressure screw. To regulate the opening and closing of the valve, take the pointed end of a file and apply it to the teeth of the regulator. If valve closes with too much boiler pressure, move the regulator to the left.   If with too little, move the regulator to the right.

This can be done when the valve is at the point of blowing off.

PHANTOM VIEW OF MARSH INDEPENDENT STEAM PUMP.

Other types of valves are managed in a similar way, and exact directions will always be furnished by the manufacturers.

FILLING THE BOILER WITH WATER.

There are three ways in which a boiler is commonly filled with water.

First, before starting a boiler it must be filled with water by hand, or with a hand force-pump. There is usually a filler plug, which must be taken out, and a funnel can be attached in its place. Open one of the gauge cocks to let out the air as the water goes in.

When the boiler has a sufficient amount of water, as may be seen by the glass water gauge, replace the filler plug. After steam is up the boiler should be supplied with water by a pump or injector.

THE BOILER PUMP.

There are two kinds of pumps commonly used on traction engines, the Independent pump, and the Crosshead pump.

The Independent pump is virtually an independent engine with pump attached. There are two cylinders, one receiving steam and conveying force to the piston; the other a water cylinder, in which a plunger works, drawing the water into itself by suction and forcing it out through the connection pipe into the boiler by force of steam pressure in the steam cylinder.

STRAIGHT GLOBE VALVE.

It is to be noted that all suction pumps receive their water by reason of the pressure of the atmosphere on the surface of the water in the supply tank or well. This atmospheric pressure is about 15 pounds to the square inch, and is sufficient to support a column of water 28 to 33 feet high, 33 feet being the height of a column of water which the atmosphere will support theoretically at about sea level. At greater altitudes the pressure of the atmosphere decreases. Pumps do not work very well when drawing water from a depth over 20 or 22 feet.

Water can be forced to almost any height by pressure of steam on the plunger, and it is taken from deep wells by deep well pumps, which suck the water 20 to 25 feet, and force it the rest of the way by pressure on a plunger.

ANGLE GLOBE VALVE.

The amount of water pumped is regulated by a cock or globe valve in the suction pipe.

A Cross-head boiler pump is a pump attached to the cross-head of an engine. The force of the engine piston is transmitted to the plunger of the pump.

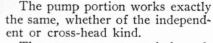

The pump portion works exactly the same, whether of the independent or cross-head kind.

The cut represents an independent pump that uses the exhaust steam to heat the water as it is pumped (Marsh pump).

Every boiler feed-pump must have at least two check valves.

A check valve is a small swinging gate valve (usually) contained in a pipe, and so arranged that when water is flowing in one direction the valve will automatically open to let the water pass, while if water should be forced in the other direction, the valve will automatically close tight and prevent the water from passing.

VALVE WITH INTERNAL SCREW.

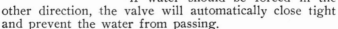

There is one check valve in the supply pipe which conducts the water from the tank or well to the pump cylinder. When the plunger is drawn back or raised, a vacuum is created in the pump cylinder and the outside atmospheric pressure forces water through the supply pipe into the cylinder, and the check valve opens to let it pass. When the plunger returns, the check valve closes, and the water is forced into the feed-pipe to the boiler.

SECTIONAL VIEW OF SWING CHECK VALVE.

There are usually two check valves between the pump cylinder and the boiler, both swinging away from the pump or toward the boiler. In order that the water may flow steadily into the boiler there is an air chamber, which may be partly filled with water at each stroke of the plunger. As the water comes in, the air must be compressed, and as it expands it forces the water through the feed pipe into the boiler in a steady stream. There is one

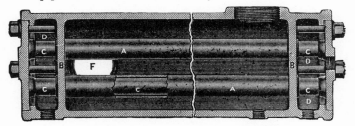

SECTIONAL VIEW OF CASE HEATER.

check valve between the pump cylinder and the air chamber, to prevent the water from coming back into the cylinder, and another between the air chamber and the boiler, to prevent the steam pressure forcing itself or the water from the boiler or water heater back into the air chamber.

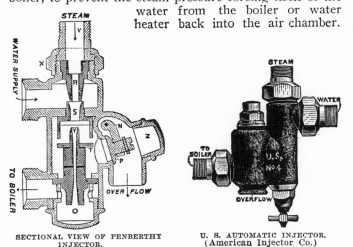

SECTIONAL VIEW OF PENBERTHY INJECTOR.

U. S. AUTOMATIC INJECTOR. (American Injector Co.)

All three of these check valves must work easily and fit tight if the pump is to be serviceable. They usually close with rubber facings which in time will get worn, and dirt is liable to work into the hinge and otherwise prevent tight and easy closing. They can always be opened for inspection, and new ones can be put in when the old are too much worn.

Only cold water can be pumped successfully, as steam from hot water will expand, and so prevent a vacuum being formed. Thus no suction will take place to draw the water from the supply source.

There should always be a globe valve or cock in the feed pipe near the boiler to make it possible to cut out the check valves when the boiler is under pressure. *It is never to be closed except* when required for this purpose.

AUTOMATIC INJECTOR.

Before passing into the boiler the water from the pump goes through the *heater*. This is a small cylinder, with a coil of pipe inside. The feed pipe from the pump is connected with one end of this inner coil of pipe, while the other end of the coil leads into the boiler itself. The exhaust steam from the engine cylinder is admitted into the cylinder and passes around the coil of pipe, afterwards coming out of the smoke stack to help increase the draft. As the feed water passes through this heater it becomes heated nearly to boiling before it enters the boiler, and has no tendency to cool the boiler off. Heating the feed water results in an economy of about 10 per cent.

The Injector is another means of forcing water from a supply tank or well into the boiler, and at the same time heating it, by use of steam from the boiler. It is a necessity when a cross-head pump is used, since such a pump will not work when the engine is shut down. It is useful in any case to heat the water before it goes into the boiler when the engine is not working and there is no exhaust steam for the heater.

There are various types of injectors, but they all work on practically the same principle. The steam from the boiler is led through a tapering nozzle to a small chamber into which there is an opening from a water supply pipe. This steam nozzle throws out its spray with great force and creates a partial vacuum in the chamber, causing the water to flow in. As the pressure of the steam has been reduced when it passes into the injector, it cannot, of course, force its way back into the boiler at first, and finds an outlet at the overflow. When the water comes in, however, the steam jet strikes the water and is condensed by it. At the same time it carries the water and the condensed steam along toward the boiler with such force that the back pressure of the boiler is overcome and a stream of heated water is passed into it. In order that the injector may work, its parts must be nicely adjusted, and with varying steam pressures it takes some ingenuity to get it started. Usually the full steam pressure is turned on and the cock admitting the water supply is opened a varying amount according to the pressure.

First the valve between the check valve and the boiler should be opened, so that the feed water may enter freely; then open wide the valve next the steam dome, and any other valve between the steam supply pipe and the injector; lastly open the water supply valve. If water appears at the overflow, close the supply valve and open it again, giving it just the proper amount of turn. The injector is regulated by the amount of water admitted.

In setting up an injector of any type, the following rules should be observed:

PLAIN WHISTLE.

All connecting pipes as straight and short as possible.

The internal diameter of all connecting pipes should be the same or greater than the diameter of the hole in the corresponding part of the injector.

When there is dirt or particles of wood or other material in the source of water supply, the end of the water supply pipe should be provided with a strainer. Indeed, invariably a strainer should be used. The holes in this strainer must be as small as the smallest opening in the delivery tube, and the total area of the openings in the strainer must be much greater than the area of the water supply (cross-section).

The steam should be taken from the highest part of the dome, to avoid carrying any water from the boiler over with it. Wet steam cuts and grooves the steam nozzle. The steam should not be taken from the pipe leading to the engine unless the pipe is quite large.

Before using new injectors, after they are fitted to the boiler it is advisable to disconnect them and clean them out well by letting steam blow through them or forcing water through. This will prevent lead or loose scale getting into the injector when in use.

Set the injector as low as possible, as it works best with smallest possible lift.

Ejectors and jet pumps are used for lifting and forcing water by steam pressure, and are employed in filling tanks, etc.

BLAST AND BLOW-OFF DEVICES.

In traction engines there is small pipe with a valve, leading into the smoke stack from the boiler. When the valve is opened, the steam allowed to blow off into the smoke stack will create a vacuum and so increase the draft. Blast or blow pipes are used only in starting the fire, and are of little value before the steam pressure reaches 15 pounds or so.

The exhaust nozzle from the engine cylinder also leads into the smoke stack, and when the engine is running the exhaust steam is sufficient to keep up the draft without using the blower.

Blow-off cocks are used for blowing sediment out of the bottom of a boiler, or blowing scum off the top of the water to prevent foaming. A boiler should never be blown out at high pressure, as there is great danger of injuring it. Better let the boiler cool off somewhat before blowing off.

DIAMOND
SPARK ARRESTER.

SPARK ARRESTER.

Traction engines are supplied as a usual thing with spark arresters if they burn wood or straw. Coal sparks are heavy and have little life, and with some engines no spark arrester is needed. But there is great danger of setting a fire if an engine is run with wood or straw without the spark arrester.

Spark arresters are of different types. The most usual form is a large screen dome placed over the top of the stack. This screen must be kept well cleaned by brushing, or the draft of the engine will be impaired by it.

In another form of spark arrester, the smoke is made to pass through water, which effectually kills every possible spark.

The Diamond Spark Arrester does not interfere with the draft and is so constructed that all sparks are carried by a counter current through a tube into a pail where water is kept. The inverted cone, as shown in cut, is made of steel wire cloth, which permits smoke and gas to escape, but no sparks. There is no possible chance to set fire to anything by sparks. It is adapted to any steam engine that exhausts into the smoke stack.

CHAPTER III.

THE SIMPLE ENGINE.

The engine is the part of a power plant which converts steam pressure into power in such form that it can do work. Properly speaking, the engine has nothing to do with generating steam. That is done exclusively in the boiler, which has already been described.

The steam engine was invented by James Watt, in

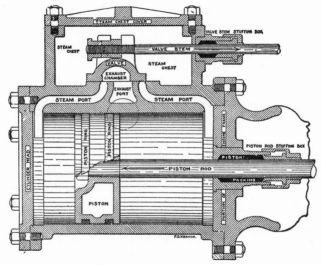

VIEW OF SIMPLE CYLINDER.
(J. I. Case Threshing Machine Co.)

England, between 1765 and 1790, and he understood all the essential parts of the engine as now built. It was improved, however, by Seguin, Ericsson, Stephenson, Fulton, and many others.

Let us first consider:

THE STEAM CYLINDER, ITS PARTS AND CONNECTIONS.

The cylinder proper is constructed of a single piece of cast iron bored out smooth.

The *cylinder heads* are the flat discs or caps bolted to the ends of the cylinder itself. Sometimes one cylinder head is cast in the same piece with the engine frame.

The *piston* is a circular disc working back and forth in the cylinder. It is usually a hollow casting, and to make it fit the cylinder steam tight, it is supplied on its circumference with *piston rings.* These are made of slightly larger diameter than the piston, and serve as springs against the sides of the cylinder. The *follower*

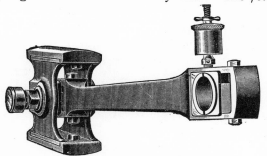

CONNECTING ROD AND CROSS-HEAD.
(J. I. Case Threshing Machine Co.)

plate and bolts cover the piston rings on the piston head and hold them in place.

The *piston rod* is of wrought iron or steel, and is fitted firmly and rigidly into the piston at one end. It runs from the piston through one head of the cylinder, passing through a steam-tight "stuffing box." One end of the piston rod is attached to the cross-head.

The *cross-head* works between *guides,* and has *shoes* above and below. It is practically a joint, necessary in converting straight back and forth motion into rotary. The cross-head itself works straight back and forth, just as the piston does, which is fastened firmly to one end. At the other end is attached the *connecting rod,* which works on a bearing in the cross-head, called the *wrist pin,* or cross-head pin.

The *connecting rod* is wrought iron or steel, working at one end on the bearing known as the wrist pin, and on the other on a bearing called the *crank pin.*

The *crank* is a short lever which transmits the power from the connecting rod to the *crank shaft.* It may also be a disc, called the *crank disc.*

Let us now return to the steam cylinder itself.

The steam leaves the boiler through a pipe leading from the top of the steam dome, and is let on or cut off by the *throttle* valve, which is usually opened and closed by some sort of lever handle. It passes on to the

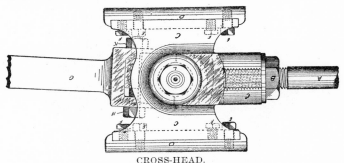

CROSS-HEAD.
(J. I. Case Threshing Machine Co.)

Steam-chest, usually a part of the same casting as the cylinder. It has a cover called the *steam-chest cover,* which is securely bolted in place.

The *steam valve,* usually spoken of simply as the *valve,* serves to admit the steam alternately to each end of the cylinder in such a manner that it works the piston back and forth.

There are many kinds of valves, the simplest (shown in the diagram) being the D-valve. It slides back and forth on the bottom of the steam-chest, which is called the *valve seat,* and alternately opens and closes the two *steam ports,* which are long, narrow passages through which the steam enters the cylinder, first through one port to one end, then through the other port to the other end. The exhaust steam also passes out at these same ports.

The *exhaust chamber* in the type of engine now under consideration is an opening on the lower side of the valve, and is always open into the *exhaust port,* which connects with the exhaust pipe, which finally discharges itself through the *exhaust nozzle* into the smoke stack of a locomotive or traction engine, or in other types of engines, into the *condenser.*

The valve is worked by the *valve stem,* which works through the valve stem *stuffing-box.*

Of course the piston does not work quite the full length of the cylinder, else it would pound against the cylinder heads.

The *clearance* is the distance between the cylinder head at either end and the piston when the piston has reached the limit of its stroke in that direction.

In most engines the valve is so set that it opens a trifle just before the piston reaches the limit of its movement in either direction, thus letting some steam in before the piston is ready to move back. This opening, which usually amounts to 1-32 to 3-16 of an inch, is called the *lead.* The steam thus let in before the piston reaches the limit of its stroke forms *cushion,* and helps the piston to reverse its motion without any jar, in an easy and silent manner. Of course the cushion must be as slight as possible and serve its purpose, else it will tend to stop the engine, and result in loss of energy. Some engines have no lead.

Setting a valve is adjusting it on its seat so that the lead will be equal at both ends and sufficient for the needs of the engine. By shortening the movement of the valve back and forth, the lead can be increased or diminished. This is usually effected by changing the eccentric or valve gear.

The *lap* of a slide valve is the distance it extends over the edges of the ports when it is at the middle of its travel.

Lap on the steam side is called outside lap; lap on the exhaust side is called inside lap. The object of lap is to secure the benefit of working steam expansively. Having lap, the valve closes one steam port before the other is opened, and before the piston has reached the end of its stroke; also of course before the exhaust is opened. Thus for a short time the steam that has been let into the cylinder to drive the piston is shut up with neither inlet nor outlet, and it drives the piston by its own expansive force. When it passes out at the exhaust it has a considerably reduced pressure, and less of its force is wasted.

Let us now consider the

VALVE GEAR.

The mechanism by which the valve is opened and closed is somewhat complicated, as various things are accomplished by it besides simply opening and closing the valve. If an engine has a *reverse lever,* it works through the valve gear; and the *governor* which regulates the speed of the engine may also operate through the valve gear. It is therefore very important.

The simplest valve gear depends for its action on a fixed eccentric.

An *eccentric* consists of a central disc called the *sheave,* keyed to the main shaft at a point to one side of its true center, and a grooved ring or *strap* surrounding it and sliding loosely around it. The strap is usually made of brass or some anti-friction metal. It is in two parts, which are bolted together so that they can be tightened up as the strap wears.

The *eccentric rod* is either bolted to the strap or forms a single piece with it, and this rod transmits its motion to the valve.

It will be seen, therefore, that the eccentric is nothing more than a sort of disc crank, which, however, does not need to be attached to the end of a shaft in the manner of an ordinary crank.

The distance between the center of the eccentric sheave and the center of the shaft is called the *throw* of the eccentric or the *eccentricity.*

The eccentric usually conveys its force through a connecting rod to the valve stem, which moves the valve.

The first modification of the simple eccentric **valve** gear is

THE REVERSING GEAR.

It is very desirable to control the movement of the steam valve, so that if desired the engine may be run in the opposite direction; or the steam force may be brought to bear to stop the engine quickly; or the travel of the valve regulated so that it will let into the cylinder only as much steam as is needed to run the engine when the load is light and the steam pressure in the boiler high.

There is a great variety of reversing gears; but **we** will consider one of the commonest and simplest first.

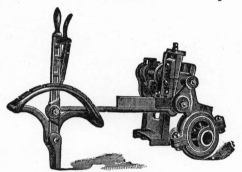

HUBER SINGLE ECCENTRIC REVERSE.

If the eccentric sheave could be slipped around on the shaft to a position opposite to that in which it was keyed to shaft in its ordinary motion, the motion of the valve would be reversed, and it would let steam in front of the advancing end of the piston, which would check its movement, and start it in the opposite direction.

The *link gear,* invented by Stephenson, accomplishes this in a natural and easy manner. There are two eccentrics placed just opposite to each other on the crank shaft, their connecting rods terminating in what is called a *link,* through which motion is communicated to the valve stem. The link is a curved slide, one eccentric being connected to one end, the other eccentric to the other end, and the *link-block,* through which motion is conveyed to the valve, slides freely from one end to the other.   Lower the link so that the block is opposite the end of the first rod, and the valve will be moved by the corresponding eccentric; raise the link, so that the block is opposite the end of the other rod, and the valve will be moved by the other eccentric. In the middle there would

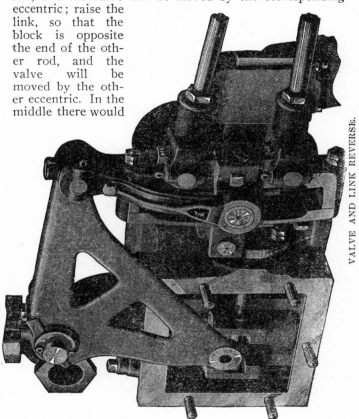

VALVE AND LINK REVERSE.

be a dead center, and if the block stopped here, the valve would not move at all.   At any intermediate point, the travel of the valve would be correspondingly shortened.

Such is he theoretical effect of a perfect link; but the dead center is not absolute, and the motion of the link is varied by the point at which the rod is attached which lifts and lowers it, and also by the length of this rod. In full gear the block is not allowed to come quite to the end of the link, and this surplus distance is called the *clearance.* The *radius* of a link is the distance from the center of the driving shaft to the center of the link, and the curve of the link is that of a circle with that radius. The length of the radius may vary considerably, but the point of suspension is important. If a link is suspended by its center, it will certainly cut off steam sooner in the front stroke than in the back. Usually it is suspended from that point which is most used in running the engine.

The *Woolf reversing gear* employs but one eccentric, to the strap of which is cast an arm having a block pivoted at its end. This block slides in a pivoted guide, the angle of which is con-

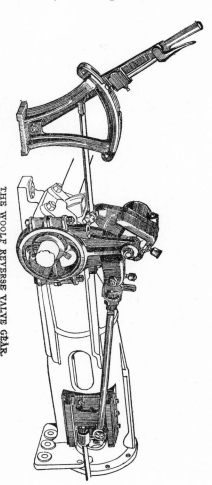

THE WOOLF REVERSE VALVE GEAR.

trolled by the reverse lever. To the eccentric arm is attached the eccentric rod, which transmits the motion to
the valve rod through a rocker arm on simple engines
and through a slide, as shown in cut, on compound engines.

The Meyer valve gear does not actually reverse an
engine, but controls the admission of steam by means
of an additional valve riding on the back of the main
valve and controlling the cut-off. The main valve is like
an ordinary D-valve, except that the steam is not admitted around the ends, but through ports running
through the valve, these ports being partially opened or
closed by the motion of
the riding valve, which is
controlled by a separate
eccentric. If this riding
valve is connected with a
governor, it will regulate
the speed of an engine;
and by the addition of a
link the gear may be
made reversible. The
chief objection to it is
the excessive friction
of the valves on their
seats.

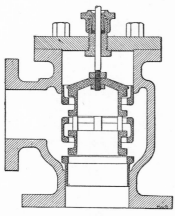

SECTIONAL VIEW SHOWING VALVE
OF WATERS GOVERNOR.

GOVERNORS.

A governor is a mech-
anism by which the sup-
ply of steam to the cylin-
der is regulated by revolving balls, or the like, which
runs faster or slower as the speed of the engine increases
or diminishes. Thus the speed of an engine is regulated
to varying loads and conditions.

The simplest type of governor, and the one commonly
used on traction engines, is that which is only a modification of the one invented by Watt. Two balls revolve
around a spindle in such a way as to rise when the speed
of the engine is high, and fall when it is low, and in rising

and falling they open and close a valve similar to the throttle valve. The amount that the governor valve is opened or closed by the rise and fall of the governor balls is usually regulated by a thumb screw at the top or side, or by what is called a handle nut, which is usually held firm by a check nut directly over it, which should be screwed firm against the handle nut.

Motion is conveyed to the governor balls by a belt and a band wheel working on a mechanism of metred cogs.

There is considerable friction about a governor of this type and much energy is wasted in keeping it going. The valve stem or spindle passes through a steam-tight stuffing box, where it is liable to stick if the packing is too tight; and if this stuffing box leaks steam, there will be immediate loss of power.

Such a governor as has just been described is called a throttle valve governor. On high grade engines the difficulties inherent in this type of governor are overcome by making the governor control, not a valve in the steam supply pipe, but the admission

PICKERING HORIZONTAL GOVERNOR.

of steam to the steam cylinder through the steam valve and its gear. Such engines are described as having an "automatic cut-off." Sometimes the governor is attached to the link, sometimes to a separate valve, as in the Meyer gear already described. Usually the governor is attached to the fly-wheel, and consequently governors of this type are called fly-wheel governors. An automatic cut-off governor is from 15 per cent to 20 per cent more effective than a throttle valve governor.

CRANK, SHAFT AND JOURNALS.

We have already seen how the piston conveys its power through the piston rod, the cross-head, and the connecting rod, to the crank pin and crank, and hence to the shaft.

The key, gib, and strap are the effective means by which the connecting rod is attached, first to the wrist pin in the cross-head, and secondly to the crank pin on the crank.

The *strap* is usually made of two or three pieces of wrought iron or steel bolted together so as to hold the *brasses*, which are in two parts and loosely surround the pin. The brasses do not quite meet, and as they wear may be tightened up. This is effected by the *gib*, back of which is the *key*, which is commonly a wedge which may be driven in, or a screw, which presses on the back of the gib, which in turn forces together the brasses; and

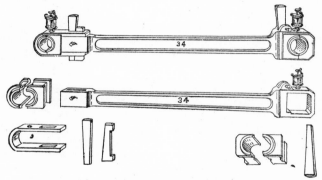

CONNECTING ROD AND BOXES.
(A. W. Stevens Co.)

thus the length of the piston gear is kept uniform in spite of the wear, becoming neither shorter nor longer. When the brasses are so worn that they have been forced together, they must be taken out and filed equally on all four of the meeting ends, and shims, or thin pieces of sheet iron or the like placed back of them to equalize the wear, and prevent the piston gear from being shortened or otherwise altered.

The *crank* is a simple lever attached to the shaft by which the shaft is rotated. There are two types of crank in common use, the side crank, which works by what is virtually a bend in the shaft. There is also what is called the disc crank, a variation of the side crank, in which the power is applied to the circumference of a disc instead of to the end of a lever arm.

The *boss* of a crank is that part which surrounds the shaft and butts against the main bearing, and is usually about twice the diameter of the crank shaft journal. The *web* of the crank is the portion between the shaft and the pin.

To secure noiseless running, the crank pin should be turned with great exactness, and should be set exactly parallel with the direction of the shaft. When the pressure on the pin or any bearing is over 800 pounds per square inch, oil is no longer able to lubricate it properly. Hence the bearing surface should always be large enough to prevent a greater pressure than 800 pounds to the square inch. To secure the proper proportions the crank pin should have a diameter of one-fourth the bore of the cylinder, and its length should be one-third that of the cylinder.

The *shaft* is made of wrought iron or steel, and must not only be able to withstand the twisting motion of the crank, but the bending force of the engine stroke. To prevent bending, the shaft should have a bearing as near the crank as possible.

The *journals* are those portions of the shaft which work in bearings. The main bearings are also called *pedestals, pillow blocks,* and *journal boxes.* They usually consist of boxes made of brass or some other anti-friction material carried in iron pedestals. The pillow blocks are usually adjustable.

THE FLY-WHEEL.

This is a heavy wheel attached to the shaft. Its object is to regulate the variable action of the piston, and to make the motion uniform even when the load is variable. By its inertia it stores energy, which would keep the engine running for some time after the piston ceased to apply any force or power.

LUBRICATORS.

All bearings must be steadily and effectively lubricated, in order to remove friction as far as possible, or the working power of the engine will be greatly reduced. Besides, without complete and effective lubrication, the bearings will "cut," or wear in irregular grooves, etc., quickly ruining the engine.

Bearings are lubricated through automatic lubricator cups, which hold oil or grease and discharge it uniformly upon the bearing through a suitable hole.

A sight feed ordinary cup permits the drops of oil to be seen as they pass downward through a glass tube, and

DESCRIPTION.

C 1—Body or Oil Reservoir.
C 3—Filler Plug.
C 4—Water Valve.
C 5—Plug for inserting Sight-
        Feed Glass.
C 6—Sight-Feed Drain Stem.
C 7—Regulating Valve.
C 8—Drain Valve.
C 9—Steam Valve.
C 10—Union Nut.
C 11—Tail Piece.
    H—Sight-Feed Glass.

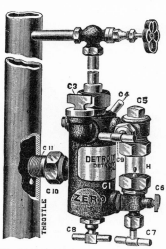

THE "DETROIT" ZERO DOUBLE CONNECTION LUBRICATOR.

also the engineer may see how much oil there is in the cup. Such a cup is suitable for all parts of an engine except the crank pin, cross-head, and, of course, the cylinder.

The crank pin oiler is an oil cup so arranged as to force oil into the bearing only when the engine is working, and more rapidly as the engine works more rapidly. In one form, which uses liquid oil, the oil stands below a disc; from which is the opening through the shank to the bearing. As the engine speeds up, the centrifugal force

tends to force the oil to the top of the cup and so on to the
bearing, and the higher the speed the greater the amount
of oil thrown into the crank pin.

Hard oil or grease has of late been coming into exten-
sive use. It is placed in a compression cup, at the top of
which a disc is pressed down by a spring, and also by some
kind of a screw. From time to time the screw is tightened up by hand, and the spring automatically forces
down the grease.

The Cylinder Lubricator is constructed on a different
principle, and uses an entirely different kind of oil, called

GLASS OIL CUP.

SECTIONAL VIEW IDEAL
GREASE CUP.

"cylinder oil." A sight-feed automatic oiler is so arranged that the oil passes through water drop by drop, so
that each drop can be seen behind glass before it passes
into the steam pipe leading from the boiler to the cylinder. The oil mingles with the steam and passes into the
steam chest, and thence into the cylinder, lubricating the
valve and piston.

The discharge of the oil may not only be watched, but
regulated, and some judgment is necessary to make sure
that enough oil is passing into the cylinder to prevent it
from cutting.

The oil is forced into the steam by the weight of the

column of water, since the steam pressure is the same at both ends. There is a small cock by which this water of condensation may be drained off when the engine is shut down in cold weather. Oilers are also injured by strain-ing from heating caused by the steam acting on cold oil when all the cocks are closed. There is a relief cock to prevent this strain, and it should be slightly opened, except when oiler is being filled.

There are a number of different types of oilers, with their cocks arranged in different ways; but the manufacturer always gives diagrams and instructions fully explaining the working of the oiler. Oil pumps serving the same purpose are now often used.

ACORN OIL PUMP.

DIFFERENTIAL GEAR.

The gearing by which the traction wheels of a traction engine are made to drive the engine is an important item. Of course, it is desirable to apply the power of the engine to both traction wheels; yet if both hind wheels were geared stiff, the engine could not turn from 'a straight line, since in turning one wheel must move faster than the other. The differential or compensating gear is a device to leave both wheels free to move one ahead of the other if occasion requires. The principle is much the same as in case of a rachet on a geared wheel, if power were applied to the ratchet to make the wheel turn; if for any reason the wheel had a tendency of its own to turn faster than the ratchet forced it, it would be free to do so. When corners are turned the power is applied to one wheel only, and the other wheel is permitted to move faster or slower than the wheel to which the gearing applies the power.

There are several forms of differential gears, differing largely as to combination of spur or bevel cogs. One of the best known uses four little beveled pinions, which are placed in the main driving wheel as shown in the cut. Beveled cogs work into these on either side of the main

wheel. If one traction wheel moves faster than the other
these pinions move around and adjust the gears on either
side.

FRICTION CLUTCH.

The power of an engine is usually applied to the traction wheel by a friction clutch working on the inside of

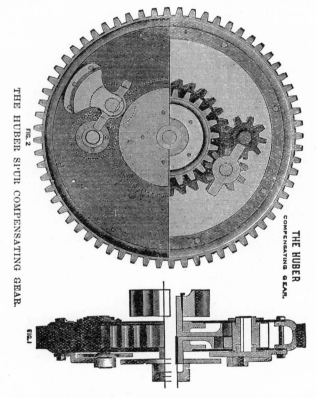

the fly-wheel. (See plan of Frick Engine.) The
traction wheels are the two large, broad-rimmed hind
wheels, and are provided with projections to give them

a firm footing on the road. Traction engines are also provided with mud shoes and wheel cleaning devices for mud and snow.

THE FUSIBLE PLUG.

The fusible plug is a simple screw plug, the center of which is bored out and subsequently filled with some other metal that will melt at a lower temperature than steel or iron. This plug is placed in the crown sheet of a locomotive boiler as a precaution for

AULTMAN & TAYLOR BEVEL COMPENSATING GEAR

safety. Should the crown sheet become free of water when the fire is very hot, the soft metal in the fusible plug would melt and run out, and the overheated steam

DIFFERENTIAL GEAR, SHOWING CUSHION SPRINGS AND BEVEL PINION.

would escape into the firebox, putting out the fire and giving the boiler relief so that an explosion would be avoided. In some states a fusible plug is required by law, and one is found in nearly every boiler which has a crown sheet. Return flue boilers and others which do not have crown sheets (as for example the vertical) do not have fusible plugs. To be of value a fusible plug should be renewed or changed once a month.

STUFFING BOXES.

Any arrangement to make a steam-tight joint about a moving rod, such as a piston rod or steam valve rod, would be called a stuffing box. Usually the stuffing box gives free play to a piston rod or valve rod, without allowing any steam to escape. A stuffing box is also used on a pump piston sometimes, or a compressed air piston. In all these cases it consists of an annular space around the moving rod which can be partly filled by some pliable elastic material such as hemp, cotton, rubber, or the like; and this filling is held in place and made tighter or looser by what is called a gland, which is forced into the partly filled box by screwing up a cap on the outside of the cylinder. Stuffing boxes must be repacked occasionally, since the packing material will get hard and dead, and will either leak steam or cut the rod.

CYLINDER COCKS.

These cocks are for the purpose of drawing the water formed by condensation of steam out of the cylinder. They should be opened whenever the engine is stopped or started, and should be left open when the engine is shut down, especially in cold weather to prevent freezing of water and consequent damage. Attention to these cocks is very important.

These are small cocks arranged about the pump and at other places for the purpose of testing the inside action. By them it is possible to see if the pump is working properly, etc.

STEAM INDICATOR.

The steam indicator is an instrument that can be attached to either end of a steam cylinder, and will indicate the character of the steam pressure during the entire stroke of the piston. It shows clearly whether the lead is right, how much cushion there is, etc. It is very important in studying the economical use and distribution of steam, expansive force of steam, etc.

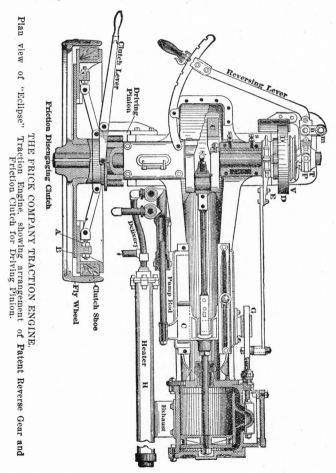

Plan view of "Eclipse" Traction Engine, showing arrangement of Patent Reverse Gear and Friction Clutch for Driving Pinion.

THE FRICK COMPANY TRACTION ENGINE.

LIST OF ATTACHMENTS FOR TRACTION ENGINE AND BOILER.

The following list of brasses, etc., which are packed with the Case traction engine will be useful for reference in connection with any similar traction engine and boiler. The young engineer should rapidly run over every new engine and locate each of these parts, which will be differently placed on different engines:

1 Steam Gauge with siphon.
1 Safety Valve.
1 Large Lubricator.
1 Small Lubricator for Pump.
1 Glass Water Gauge complete with glass and rods.
2 Gauge Cocks.
1 Whistle.
1 Injector Complete.
1 Globe Valve for Blow-off.
1 Compression Grease Cup for Cross Head.
1 Grease Cup for Crank Pin.
1 Oiler for Reverse Block.
1 Glass Oiler for Guides.
1 Small Oiler for Eccentric Rod.
1 Cylinder Cock (1 is left in place.
2 Stop Cocks to drain Heater.
1 Stop Cock for Hose Coupling on Pump.

1 Bibb Nose Cock for Pump.
1 Pet Cock for Throttle.
2 Pet Cocks for Steam Cylinder of Pump.
1 Pet Cock for Water Cylinder of Pump.
1 Pet Cock for Feed Pipe from Pump.
1 Pet Cock for Feed Pipe from Injector.
1 Governor Belt.
1 Flue Cleaner.
15 ft. lin. Suction Hose.
5 ft. Sprinkling Hose.
1 Strainer for Suction Hose.
1 Strainer for Funnel.
4 ft. 6 in. of in. Hose for Injector.
5 ft. 6 in of in. Hose for Pump.
2 Nipples ¾x2½ in. for Hose.
2 ¾ in. Hose Clamps.
2 ½ in. Hose Strainers.

TEST QUESTIONS ON BOILER AND ENGINE

Q. How is the modern stationary fire-flue boiler arranged?

Q. How does the locomotive type of boiler differ?

Q. What is a return flue boiler?

Q. What is a water-tube boiler and how does it differ from a fire-flue tubular boiler?

Q. What is a vertical boiler and what are its advantages?

Q. What is the shell?

Q. What are the boiler heads?

Q. What are the tube sheets?

Q. What is the firebox?

Q. What is the water leg?

Q. What is the crown-sheet?

Q. Where is the smoke-box located?

Q. What is the steam dome intended for?

Q. What is the mud-drum for?

Q. What are man-holes and hand-holes for?

Q. What is a boiler jacket?

Q. What is a steam jacket?

Q. Where is the ash-pit?

Q. What are dead-plates?

Q. How is grate surface measured?

Q. What is forced draft?

Q. How is heating surface measured?

Q. What is steam space?

Q. What is water space?

Q. What is a diaphragm plate?

Q. What is the first duty of an engineer in taking charge of a new boiler?

Q. What are the water gauge and try cocks for, and how are they placed?

Q. What is the steam gauge and how may it be tested?

Q. What is a safety valve? Should it be touched by the engineer? How may he test it with the steam gauge?

Q. How is a boiler first filled with water?

Q. How is it filled when under pressure?

Q. What is an independent pump? What is a cross-head pump?

Q. What is a check valve, and what is its use, and where located?

Q. What is a heater and how does it work?

Q. What is an injector, and what is the principle of its operation?

Q. Where are the blow-off cocks located? How should they be used?

Q. In what cases should spark arrester be used?

Q. Who invented the steam engine, and when?

Q. What are the essential parts of a steam engine?

Q. What is the cylinder, and how is it used?

Q. What is the piston, and how does it work? The piston-rings?

Q. What is the piston rod and how must it be fastened?

Q. What is the crosshead, and how does it move? What are guides or ways? Shoes?

Q. What is the connecting rod? Wrist pin? Crank pin?

Q. What is the crank? Crank shaft?

Q. Where is the throttle valve located, and what does opening and closing it do?

Q. What is the steam chest for, and where is it placed?

Q. What is a steam valve? Valve seats? Ports?

Q. What is the exhaust? Exhaust chamber? Exhaust port? Exhaust nozzle? What is a condenser?

Q. How is the valve worked, and what duties does it perform, and how?

Q. What is clearance?

Q. What is lead?

Q. What is cushion?

Q. How would you set a valve? What is lap?

Q. How is a steam valve moved back and forth in its seat?

Q. How may an engine be reversed?

Q. What is a governor, and how does it work?

Q. What is an eccentric? Eccentric sheave? Strap? Rod?

Q. What is the throw of an eccentric?

Q. How does the link reversing gear work?

Q. How does the Woolf reverse gear work?

Q. How does the Meyer valve gear work? Will it reverse an engine?

Q. What are the chief difficulties in the working of a governor?

Q. What are key, gib, and strap? Brasses?

Q. What is the boss of a crank? Web?

Q. How may noiseless running of a crank be secured?

Q What are journals? Pedestals? Pillow blocks? Journal boxes?

Q. What is the object in having a fly wheel?

Q. What different kinds of lubricators are there? Where may hard oil or grease be used? Is the oil used for lubricating the cylinder the same as that used for rest of engine?

Q. How does a cylinder lubricator work?

Q. What is differential gear, and what is it for?

Q. What is the use of a fusible plug, and how is it arranged?

Q. What are stuffing-boxes, and how are they constructed?

Q. What are cylinder cocks, and what are they used for?

Q. What are pet cocks?

Q. What is a steam indicator?

CHAPTER IV.

HOW TO MANAGE A TRACTION ENGINE BOILER.

We will suppose that the young engineer fully understands all parts of the boiler and engine, as explained in the preceding chapters. It is well to run over the questions several times, to make sure that every point has been fully covered and is well understood.

We will suppose that you have an engine in good running order. If you have a new engine and it starts off nice and easy (the lone engine without load) with twenty pounds steam pressure in the boiler, you may make up your mind that you have a good engine to handle and one that will give but little trouble. But if it requires fifty or sixty pounds to start it, you want to keep your eyes open, for something is tight. But don't begin taking the engine to pieces, for you might get more pieces than you know what to do with. Oil every bearing fully, and then start your engine and let it run for a while. Then notice whether you find anything getting warm. If you do, stop and loosen up a very little and start again. If the heating still continues, loosen again as before. But remember, loosen but little at a time, for a box or journal will heat from being too loose as quickly as from being too tight, and if you have found a warm box, don't let that box take all your attention, but keep your eye on the other bearings.

In the case of a new engine, the cylinder rings may be a little tight, and so more steam pressure will be required to start the engine; but this is no fault, for in a day or two they will be working all right if kept well oiled.

In starting a new engine trouble sometimes comes from the presence of a coal cinder in some of the boxes, which has worked in during shipment. Before starting a new engine, the boxes and oil holes should therefore be thoroughly cleaned out. For this purpose the engineer should always have some cotton waste or an oiled rag ready for constant use.

A new engine should be run slowly and carefully until it is found to be in perfect running order.

If you are beginning on an old engine in good running order, the above instructions will not be needed; but it is well to take note of them.

Now if your engine is all right, you may run the pressure up to the point of blowing off, which is 100 to 130 pounds, at which most safety valves are set at the factory. It is not uncommon for a new pop to stick, and as the steam runs up it is well to try it by pulling the relief lever. If on letting it go it stops the escaping steam at once, it is all right. If, however, the steam continues to escape the valve sticks in the chamber. Usually a slight tap with a wrench or hammer will stop it at once; but don't get excited if the steam continues to escape. As long as you have plenty of water in the boiler, and know that you have it, you are all right.

STARTING UP A BOILER.

Almost the only danger from explosion of a boiler is from not having sufficient water in the boiler. The boiler is filled in the first place, as has already been explained, by hand through a funnel at the filler plug, or by a force pump. The water should stand an inch and a half in the glass of the water gauge before the fire is started. It should be heated up slowly so as not to strain the boiler or connections. When the steam pressure as shown by the steam gauge is ten or fifteen pounds, the blower may be used to increase the draft.

If you let the water get above the top of the glass, you are liable to knock out a cylinder head; and if you let the water get below the bottom of the glass, you are likely to explode your boiler.

The glass gauge is not to be depended upon, however, for a number of things may happen to interfere with its working. Some one may inadvertently turn off the gauge cocks, and though the water stands at the proper height in the glass, the water in the boiler will be very different.

A properly made boiler is supplied with two to four try-cocks, one below the proper water line, and one above it. If there are more than two they will be distributed at suitable points between.

When the boiler is under pressure, turn on the lower try-cock and you should get water. You will know it because it will appear as white mist. Then try the upper try-cock, and you will get steam, which will appear blue.

NEVER FAIL TO USE THE TRY-COCKS FREQUENTLY. This is necessary not only because you never know when the glass is deceiving you; but if you fail to use them they will get stopped up with lime or mud, and when you need to use them they will not work.

In order also to keep the water gauge in proper condition, it should be frequently blown out in the following manner: Shut off the top gauge cock and open the drain cock at the bottom of the gauge. This allows the water and steam to blow through the lower cock of the water gauge, and you know that it is open. Any lime or mud that has begun to accumulate will also be carried off. After allowing the steam to escape a few seconds, shut off the lower gauge cock, and open the upper one, and allow it to blow off about the same time. Then shut the drain cock and open both gauge cocks, when you will see the water seek its level, and you can feel assured that it is reliable and in good working condition. This little operation you should perform every day you run your engine. If you do you will not *think* you have sufficient water in the boiler, but will *know*. The engineer who always *knows* he has water in the boiler will not be likely to have an explosion. Especially should you never start your fire in the morning simply because you see water in the gauge. You should *know* that there is water in the boiler.

Now if your pump and boiler are in good working condition, and you leave the globe valve in the supply pipe to the pump open, with the hose in the tank, you will probably come to your engine in the morning and find the boiler nearly full of water, and you will think some one has been tampering with the engine. The truth is, however, that as the steam condensed, a vacuum was formed, and the water flowed in on account of atmospheric pressure, just as it flows into a suction pump when the plunger rises and creates a vacuum in the pump. Check valves are arranged to prevent anything passing out of the boiler, but there is nothing to prevent water passing in.

The only other cause of an explosion, beside poor material in the manufacture of the boiler, is too high steam pressure, due to a defective safety valve or imperfect team gauge. The steam gauge is likely to get out of order in a number of ways, and so is the safety valve. To make sure that both are all right, the one should frequently be tested by the other. The lever of the safety valve should frequently be tried from time to time, to make sure the valve opens and closes easily, and whenever the safety valve blows off, the steam gauge should be noted to see if it indicates the pressure at which the safety has been set.

WHEN YOUR ENGINE IS ALL RIGHT, LET IT ALONE.

Some engineers are always loosening a nut here, tightning up a box there, adjusting this, altering that. When an engine is all right they keep at it till it is all wrong. As a result they are in trouble most of the time. When an engine is running all right, LET IT ALONE. Don't think you are not earning your salary because you are nerely sitting still and looking on. If you must be at work, keep at it with an oily rag, cleaning and polishing up. That is the way to find out if anything is really the matter. As the practised hand of the skilled engineer goes over an engine, his ears wide open for any peculiarity of sound, anything that is not as it should be will make itself decidedly apparent. On the other hand, an engineer who does not keep his engine clean and bright by constantly passing his hand over it with an oily rag, is certain to overlook something, which perhaps in the end will cost the owner a good many dollars to put right.

Says an old engineer* we know, "When I see an engineer watching his engine closely while running, I am most certain to see another commendable feature in a

---

*J. H. Maggard, author of "Rough and Tumble Engineering," to whom we are indebted for a number of valuable suggestions in this chapter.

good engineer, and that is, when he stops his engine he will pick up a greasy rag and go over his engine carefully, wiping every working part, watching or looking carefully at every point that he touches.   If a nut is working loose, he finds it; if a bearing is hot, he finds it; if any part of his engine has been cutting, he finds it.   He picks up a greasy rag instead of a wrench, for the engineer that understands his business and attends to it never picks up a wrench unless he has something to do with it."

This same engineer goes on with some more most excellent advice.   Says he:

"Now, if your engine runs irregularly, that is, if it runs up to a higher speed than you want, and then runs down, you are likely to say at once, 'Oh, I know what the trouble is, it is the governor.'   Well, suppose it is.   What are you going to do about it?   Are you going to shut down at once and go to tinkering with it?   No, don't do that.   Stay close to the throttle valve and watch the governor closely.   Keep your eye on the governor stem and when the engine starts off on one of its speed tilts, you will see the stem go down through the stuffing box and then stop and stick in one place until the engine slows down below its regular speed, and it then lets loose and goes up quickly and your engine lopes off again.   You have now located the trouble.   It is in the stuffing box around the little brass rod or governor stem.   The packing has become dry and by loosening it up and applying oil you may remedy the trouble until such time as you can repack it with fresh packing.   Candle wick is as good for this purpose as anything you can use.

"But if the governor does not act as I have described, and the stem seems to be perfectly free and easy in the box, and the governor still acts queerly, starting off and running fast for a few seconds and then suddenly concluding to take it easy and away goes the engine again, see if the governor belt is all right, and if it is it would be well for you to stop and see if a wheel is not loose.   It might be either the little belt wheel or one of the little cog wheels.   If you find these are all right, examine the spool on the crank shaft from which the governor is run, and you will probably find it loose. If the engine has been run for any length of time, you will always find the trouble in one of these places; but if it is a new one, the governor valve might work a little tight in the valve chamber, and you may have to take it out and use a little emery paper to take off the rough projections on the valve. Never use a file on this valve if you can get emery paper, and I should advise you always to have some of it with you. It will often come handy."

This is good advice in regard to any trouble you may have with an engine. Watch the affected part closely; think the matter over carefully, and see if you cannot locate the difficulty before you even stop your engine. If you find the trouble and know that you have found it, you will soon be able to correct the defect, and no time will be lost. At the same time you will not ruin your engine by trying all sorts of remedies at random in the thought that you may ultimately hit the right thing. The chances are that before you do hit the right point, you will have put half a dozen other matters wrong, and it will take half a day to get the matter right again.

As there are many different types of governors in use, it would be impossible to give exact directions for regulating that would apply to them all; but the following suggestions applying to the Waters governor (one widely used on threshing engines) will give a general idea of the method for all:

There are two little brass nuts on the top of the stem of the governor, one a thumb nut and the other a loose jam nut. To increase the speed, loosen the jam nut and then turn the thumb nut back slowly, watching the motion of the engine all the time. When the required speed has been obtained, then tighten up as snug as you can with your fingers (not using a wrench). To decrease the speed, loosen the jam nut as before, running it up a few turns, and then turn down the thumb nut till the speed meets your requirements, when the thumb nut is made fast as before. In any case, be very careful not to press down on the stem when turning the thumb nut, as this will make the engine run a little slower than will be the case when your hand has been removed.

If your engine does not start with an open throttle, look to see if the governor stem has not been screwed down tight. This is usually the case with a new engine, which has been screwed down for safety in transportation.

WATER FOR THE BOILER.

There is nothing that needs such constant watching and is likely to cause so much trouble if it is not cared for, as the supply of water. Hard well water will coat the inside of the boiler with lime and soon reduce its steaming power in a serious degree, to say nothing of stopping up pipes, cocks, etc. At the same time, rain water that is perfectly pure (theoretically) will be found to have a little acid or alkali in it that will eat through the iron or steel and do equal damage.

However, an engineer must use what water he can. He cannot have it made to order for him, but he must take it from well, from brook, or cistern, or roadside ditch, as circumstances may require. The problem for the engineer is not to get the best water, but to make the best use of whatever water he can get, always, of course, choosing the best and purest when there is such a thing as choosing.

In the first place, all supply pipes in water that is muddy or likely to have sticks, leaves, or the like in it, should be furnished with strainers. If sticks or leaves get into the valve, the expense in time and worry to get them out will be ten times the cost of a strainer.

If the water is rain water, and the boiler is a new one, it would be well to put in a little lime to give the iron a slight coating that will protect it from any acid or alkali corrosion.

If the water is hard, some compound or sal ammonia should be used. No specific directions can be given, since water is made hard by having different substances dissolved in it, and the right compound or chemical is that which is adapted to the particular substance you are to counteract. An old engineer says his advice is to use no compound at all, but to put a hatful of potatoes in the boiler every morning.

Occasionally using rain water for a day or two previous to cleaning is one of the best things in the world to remove and throw down all scale. It beats compounds at every point. It is nature's remedy for the bad effects of hard water.

The important thing, however, is to clean the boiler thoroughly and often. In no case should the lime be allowed to bake on the iron. If it gets thick, the iron or steel is sure to burn, and the lime to bake so hard it will be almost impossible to get it off. But if the boiler is cleaned often, such a thing will not happen.

Mud or sediment can be blown off by opening the valve from the mud drum or the firebox at the bottom of the boiler when the pressure is not over 15 or 20 pounds; and at this pressure much of the lime distributed about the boiler may be blown off. But this is not enough. The inside of the boiler should be scraped and thoroughly washed out with a hose and force-pump just as often as the condition of the water requires it.

In cleaning the boiler, always be careful to scrape all the lime off the top of the fusible plug.

THE PUMP.

In order to manage the pump successfully, the young engineer must understand thoroughly its construction as already described. It is also necessary to understand something of the theory of atmospheric pressure, lifting power, and forcing power.

First see that the cocks or globe valves (whichever are used) are open both between the boiler and the pump and between the pump and the water supply. The globe valve next the boiler should *never* be closed, except when examining the boiler check valve. Then open the little pet cock between the two upper horizontal check valves. Be sure that the check valves are in good order, so that water can pass only in one direction. A clear, sharp click of the check valves is certain evidence that the pump is working well. If you cannot hear the click, take a stick or pencil between your teeth at one end, put the other end on the valve, stuff your fingers in your ears, and you will hear the movement of the valve as plainly as if it were a sledge-hammer.

The small drain cock between the horizontal check valves is used to drain hot water out of the pump in starting, for a pump will never work well with hot water in it; and to drain off all water in closing down in cold weather, to prevent damage from freezing. It also assists in testing the working of the pump. In starting up it may be left open. If water flows from the drain cock, we know the pump is working all right, and then close the drain cock. If you are at any time in doubt as to whether water is going into the boiler properly, you may open this drain cock and see if cold water flows freely. If it does, everything is working as it should. If hot water appears, you may know something is wrong. Also, to test the pump, place your hand on the two check valves, and if they are cold, the pump is all right; if they are hot, something is wrong, since the heat must come from the boiler, and no hot water or steam should ever be allowed to pass from the boiler back to the pump.

A stop cock next the boiler is decidedly preferable to a globe valve, since you can tell if it is open by simply looking at it; whereas you must put your hand on a globe valve and turn it. Trouble often arises through inadvertently closing the valve or cock next the boiler, in which case, of course, no water can pass into the boiler, and the pump is likely to be ruined, since the water must get out somewhere. Some part of the pump would be sure to burst if worked against a closed boiler cock or valve.

Should the pump suddenly cease to work or stop, first see if you have any water in the tank. If there is water, stoppage may be due to air in the pump chamber, which can get in only through the stuffing-box. If this is true, tighten up the pump plunger stuffing-box nut a little. If now the pump starts off well, you have found the difficulty; but at the first opportunity you ought to repack the stuffing-box.

If the stuffing-box is all right, examine the supply suction hose. See that nothing is clogging the strainer, and ascertain whether the water is sucked in or not. If it is sucked in and then is forced out again (which you can ascertain by holding your hand lightly over the suction pipe), you may know something is the matter with the first check valve. Probably a stick or stone has gotten into it and prevents it from shutting down.

If there is no suction, examine the second check valve. If there is something under it that prevents its closing, the water will flow back into the pump chamber again as soon as the plunger is drawn back.

You can always tell whether the trouble is in the second check or in the hot water check valve by opening the little drain cock. If hot water flows from it, you may know that the hot water check valve is out of order; if only cold water flows, you may be pretty sure the hot water check is all right. If there is any reason to suspect the hot water check valve, close the stop cock or valve next the boiler before you touch the check in any way. To tamper with the hot water check while the steam pressure is upon it would be highly dangerous, for you are liable to get badly burned with escaping steam or hot water. At the same time, be very sure the stop cock or valve next the boiler is open again before you start the pump.

Another reason for check valves refusing to work besides having something under them, is that the valve may stick in the valve chamber because of a rough place in the chamber, or a little projection on the valve. Light tapping with a wrench may remedy the matter. If that does not work, try the following plan suggested by an old engineer*: "Take the valve out, bore a hole in a board about one-half inch deep, and large enough to permit the valve to be turned. Drop a little emery dust in this hole If you haven't any emery dust, scrape some grit from a whetstone. If you have no whetstone, put some fine sand or gritty soil in the hole, put the valve on top of it, put your brace on the valve and turn it vigorously for a few minutes, and you will remove all roughness."

*J. H Maggard.

Sometimes the burr on the valve comes from long use; but the above treatment will make it as good as new.

INJECTORS.

All injectors are greatly affected by conditions, such as the lift, the steam pressure, the temperature of the water, etc. An injector will not use hot water well, if at all. As the lift is greater, the steam pressure required to start is greater, and at the same time the highest steam pressure under which the injector will work at all is greatly decreased. The same applies to the lifting of warm water: the higher the temperature, the greater the steam pressure required to start, and the less the steam pressure which can be used as a maximum.

It is important for the sake of economy to use the right sized injector. Before buying a new injector, find out first how much water you need for your boiler, and then buy an injector of about the capacity required, though of course an injector must always have a maximum capacity in excess of what will be required.

If the feed water is cold, a good injector ought to start with 25 pounds steam pressure and work up to 150 pounds for a 2-foot lift. If the lift is eight feet, it will start at 30 pounds and work up to 130. If the water is heated to 100 degrees Fahrenheit it will start for a 2-foot lift with 26 pounds and work up to 120 pounds, or for an 8-foot lift, it will start with 33 pounds and work up to 100. These figures apply to the single tube injector. The double tube injector should work from 14 pounds to 250, and from 15 to 210 under same conditions as above. The double tube injector is not commonly used on farm engines, however.

Care should be taken that the injector is not so near the boiler as to become heated, else it will not work. If it gets too hot, it must be cooled by pouring cold water on the outside, first having covered it with a cloth to hold the water. If the injector is cool, and the steam pressure and lift are all right, and still the injector does not work, you may be sure there is some obstruction somewhere. Shut off the steam from the boiler, and run a fine wire down through the cone valve or cylinder valve, after having removed the cap or plug nut.

Starting an injector always requires some skill, and injectors differ.   Some start by manipulating the steam valve; some require that the steam be turned on first, and then the water turned on in just the right amount, usually with a quick short twist of the supply valve.   Often some patience is required to get just the right turn on it so that it will start.

Of course you must be sure that all joints are air-tight, else the injector will not work under any conditions.

Never use an injector where a pump can be used, as the injector is much more wasteful of steam.   It is for an emergency or to throw water in a boiler when engine is not running.

No lubricator is needed on an injector.

THE HEATER.

The construction of the heater has already been explained.   It has two check valves, one on the side of the pump and one on the side of the boiler, both opening toward the boiler.   The exhaust steam is usually at a temperature of 215 to 220 degrees when it enters the heater chamber, and heats the water nearly or quite to boiling point as it passes through.   The injector heats the water almost as hot.

The heater requires little attention, and the check valves seldom get out of order.

The pump is to be used when the engine is running, and the injector when the engine is closed down.   The pump is the more economical; but when the engine is not working the exhaust steam is not sufficient to heat the water in the heater; and pumping cold water into the boiler will quickly bring down the pressure and injure the boiler.

ECONOMICAL FIRING.

The management of the fire is one of the most important things in running a steam engine.   On it depend two things of the greatest consequence—success in getting up steam quickly and keeping it at a steady pressure under all conditions; and economy in the use of fuel. An engineer who understands firing in the most economical way will probably save his wages to his employer over the engineer who is indifferent or unscientific about it. Therefore the young engineer should give the subject great attention.

First, let us consider firing with coal. All expert engineers advise a "thin" fire. This means that you should have a thin bed of coals, say about four inches thick, all over the grate. There should be no holes or dead places in this, for if there are any, cold air will short-circuit into the fire flues and cool off the boiler.

The best way of firing is to spread the coal on with a small hand shovel, a very little at a time, scattering it well over the fire. Another way, recommended by some, is to have a small pile of fresh fuel at the front of the grate, pushing it back over the grate when it is well lighted. To manage this well will require some practice and skill, and for a beginner, we recommend scattering small shovelsful all over the fire. All lump coal should be broken to a uniform size. No piece larger than a man's fist should be put in a firebox.

Seldom use the poker above the fire, for nothing has such a tendency to put out a coal fire as stirring it with a poker above. And when there is a good glow all over the grate below, the poker is not needed below. When the grate becomes covered with dead ashes, they should be cautiously but fully removed, and clinkers must be lifted out with the poker from above, care being exercised to cover up the holes with live coals.

Hard coal if used should be dampened before being put on the fire.

When the fire is burning a little too briskly, close the draft but do not tamper with the fire itself. Should it become important on a sudden emergency to check the fire at any time quickly, never dash water upon it, but rather throw plenty of fresh fuel upon it. Fresh fuel always lowers the heat at first. If all drafts are closed tight, it will lower the heat considerably for quite a time.

In checking a fire, it must be remembered that very

sudden cooling will almost surely crack the boiler. If there is danger of an explosion it may be necessary to draw the fire out entirely; but under no circumstances should cold water be thrown on. After drawing the fire close all doors and dampers.

FIRING WITH WOOD.

Always keep the fire door shut as much as possible, as cold air thus admitted will check the fire and ruin the boiler.

Firing with wood is in many ways the exact reverse of firing with coal. The firebox should be filled full of wood at all times. The wood should be thrown in in every direction, in pieces of moderate size, and as it burns away, fresh pieces should be put in at the front so that they will get lighted and ready to burn before being pushed back near the boiler. It often helps a wood fire, too, to stir it with a poker. Wood makes much less ash than coal, and what little accumulates in the grate will not do much harm. Sometimes green wood will not burn because it gets too much cold air. In that case the sticks should be packed as close together as possible, still leaving a place for the air to pass. Also a wood fire, especially one with green wood, should be kept up to a high temperature all the time; for if it is allowed to drop down the wood will suddenly cease to burn at all.

FIRING WITH STRAW.

In firing with straw it is important to keep the shute full of straw all the time so that no cold air can get in on top of the fire. Don't push the straw in too fast, either, but keep it moving at a uniform rate, with small forkfulls. Now and then it is well to turn the fork over and run it down into the fire to keep the fire level. Ashes may be allowed to fill up in rear of ash box, but fifteen inches should be kept clear in front to provide draft. The brick arch may be watched from the side opening in the firebox, and should show a continuous stream of white flame coming over it. If too much straw is forced in, that will check the flame. The flame should never be checked. If damp straw gets against the ends of the flues, it should be scraped off with the poker from side door. Clean the tubes well once a day. The draft must always be kept strong enough to produce a white heat, and if this cannot be done otherwise, a smaller nozzle may be used on the exhaust pipe; but this should be avoided when possible, since it causes back pressure on the engine. Never let the front end of the boiler stand on low ground. Engine should be level, or front end high, if it has a firebox locomotive boiler; if a return flue boiler, be careful to keep it always level. In burning straw take particular notice that the spark screen in stack does not get filled up.

THE ASH PIT.

In burning coal it is exceedingly important that the ashes be kept cleaned out, as the hot cinders falling down on the heap of ashes almost as high as the grate will overheat the grate in a very short time and warp it all out of shape, so ruining it.

With wood and straw, on the contrary, an accumulation of ashes will often help and will seldom do any harm, because no very hot cinders can drop down below the grates, and the hottest part of the fire is some distance above the grates.

STARTING A FIRE.

You must make up your mind that it will take half an hour to an hour or so to get up steam in any boiler that is perfectly cold. The metal expands and shrinks a great deal with the heat and cold, and a sudden application of heat would ruin a boiler in a short time. Hence it is necessary for reasons of engine economy to make changes of temperature, either cooling off or heating up, gradually.

First see that there is water in the boiler.

Start a brisk fire with pine kindlings, gradually putting on coal or wood, as the case may be, and spreading the fire over the grate so that all parts will be covered with glowing coals.

When you have 15 or 20 pounds of steam, start the

blower. As has already been described, the blower is a pipe with a nozzle leading from the steam space of the boiler to the smoke stack, and fitted with a globe valve. The force of the steam drives the air out of the stack, causing a vacuum, which is immediately filled by the hot gases from the firebox coming through the boiler tubes. Little is to be gained by using the blower with less than 15 pounds of steam, as the blower has so little strength below that, that it draws off about as much steam as is made and nothing is gained.

The blower is seldom needed when the engine is working, as the exhaust steam should be sufficient to keep the fire going briskly. If it is not, you should conclude that something is the matter. There are times, however, when the blower·is required even when the engine is going. For example, if you are working with very light load and small use of steam, the exhaust may be insufficient to keep up the fire; and this will be especially true if the fuel is very poor. In such a case, turn on the blower very slightly. But remember that you are wasting steam if you can get along without the blower.

Examine the nozzle of the blower now and then to see that it does ᴎot become limed up, or turned so as to direct the steam to ᴐne side of the stack, where its force would be wasted.

Beware, also, of creating too much draft; for too much draft will use up fuel and make little steam.

SMOKE.

Coal smoke is nothing more or less than unburned carbon. The more smoke you get, the less will be the heat from a given amount of fuel. Great clouds of black smoke from an engine all the time are a very bad sign in an engineer. They show that he does not know how to fire. He has not followed the directions already given, to have a thin, hot fire, with few ashes under his grate. Instead, he throws on great shovelsful of coal at a time, and has the coal up to the firebox door. His fuel is al·ways making smoke, which soon clogs up the smoke flues and lessens the amount of steam he is getting. If he had kept his fire very "thin," but very hot, throwing on a small hand shovel of coal at a time, seldom poking his fire except to lift out clinkers or clean away dead ashes under the grate, and keeping his ashpit free from ashes, there would be only a little puff of black smoke when the fresh coal went on, and then the smoke would quickly disappear, while the fire flues would burn clean and not get clogged up with soot.

It is important, however, to keep the small fire flues especially well cleaned out with a good flue cleaner; for all accumulation of soot prevents the heat from passing through the steel, and so reduces the heating capacity of the boiler. Cleaning the tubes with a steam blower is never advisable, as it forms a paste on the tube that greatly impairs its commodity.

SPARKS.

With coal there is little danger of fires caused by sparks from the engine. What sparks there are are heavy and dead, and will even fall on a pile of straw without setting it on fire. On a very windy day, however, when you are running your engine very hard, especially if it is of the direct locomotive boiler type, you want to be careful even with coal.

With wood it is very different; and likewise with straw. Wood and straw sparks are always dangerous, and an engine should never be run for threshing with wood or straw without using a spark-arrester.

It sometimes happens that when coal is used it will give out, and you will be asked to finish your job with wood. In such a case, it is the duty of an engineer to state fully and frankly the danger of firing with wood without a spark arrester, and he should go on only when ordered to do so by the proprietor, after he has been fully warned. In that case all responsibility is shifted from the engineer to the owner.

THE FUSIBLE PLUG.

The careful engineer will never have occasion to do anything to the fusible plug except to clean the scale off from the top of it on the inside of the boiler once a week, and put in a fresh plug once a month. It is put in merely as a precaution to provide for carelessness. The engineer who allows the fusible plug to melt out is by that very fact marked as a careless man, and ought to find it so much the harder to get a job.

As has already been explained, the fusible plug is a plug filled in the middle with some metal that will melt at a comparatively low temperature. So long as it is covered with water, no amount of heat will melt it, since the water conducts the heat away from the metal and never allows it to rise above a certain temperature. When the plug is no longer covered with water, however,—in short, when the water has fallen below the danger line in the boiler— the metal in the plug will fuse, or melt, and make an opening through which the steam will blow into the firebox and put out the fire. However, if the top of the fusible plug has been allowed to become thickly coated with scale, this safety precaution may not work and the boiler may explode. In any case the fusible plug is not to be depended on.

At the same time a good engineer will take every precaution, and one of these is to keep the top of the plug well cleaned. Also he will have an extra plug all ready and filled with composition metal, to put in should the plug in the boiler melt out. Then he will refill the old plug as soon as possible. This may be done by putting a little moist clay in one end to prevent the hot metal from running through, and then pouring into the other end of the plug as much melted metal as it will hold. When cold, tamp down solidly.

LEAKY FLUES.

One common cause of leaky flues is leaving the fire door open so that currents of cold air will rush in on the heated flues and cause them, or some other parts of the boiler, to contract too suddenly. The best boiler made may be ruined in time by allowing cold currents of air to strike the heated interior. Once or twice will not do it; but continually leaving the fire door open will certainly work mischief in the end.

Of course, if flues in a new boiler leak, it is the fault of the boiler maker. The tubes were not large enough to fill the holes in the tube sheets properly. But if a boiler runs for a season or so and then the flues begin to leak, the chances are that it is due to the carelessness of the engineer. It may be he has been making his fires too hot; it may be leaving the firebox door open; it may be running the boiler at too high pressure; it may be blowing out the boiler when it is too hot; or blowing out the boiler when there is still some fire in the firebox; it may be due to lime encrusted on the inside of the tube sheets, causing them to overheat. Flues may also be made to leak by pumping cold water into the boiler when the water inside is too low; or pouring cold water into a hot boiler will do it. Some engineers blow out their boilers to clean them, and then being in a hurry to get to work, refill them while the metal is hot. The flues cannot stand this, since they are thinner than the shell of the boiler and cool much more quickly: hence they will contract much faster than the rest of the boiler and something has to come loose.

Once a flue starts to leaking, it is not likely to stop till it has been repaired; and one leaky flue will make others leak.

Now what shall you do with a leaky flue?

To repair a leaky flue you should have a flue expander and a calking tool, with a light hammer. If you are small enough you will creep in at the firebox door with a candle in your hand. First, clean off the ends of the flues and flue sheet with some cotton waste. Then force the expander into the leaky flue, bringing the shoulder well up against the end of the flue. Then drive in the tapering pin. Be very careful not to drive it in too far, for if you expand the flue too much, you will strain the flue sheet and cause other flues to leak. You must use your judgment and proceed cautiously. It is better to make two or three trials than to spoil your boiler by bad work. The roller expander is preferable to the Prosser in the hands

of a novice. The tube should be expanded only enough
to stop the leak. Farther expanding will only do injury.
When you think the flue has been expanded enough, hit
the pin a side blow to loosen it. Then turn the expander
a quarter round, and drive in the pin again. Loosen up
and continue till you have turned the expander entirely
around.

Finally remove the expander, and use the calking tool
to bead the end. It is best, however, to expand all leaky
flues before doing any beading.

The beading is done by placing the guide or gauge inside the flue, and then pounding the ends of the flue down
against the flue sheet by light blows. Be very careful not
to bruise the flue sheet or flues, and use no heavy blows,
nor even a heavy hammer. Go slowly and carefully
around the end of each flue; and if you have done your
work thoroughly and carefully the flues will be all right.
But you should test your boiler before steaming up, to
make sure that all the leaks are stopped, especially if there
have been bad ones.

There are various ways to testing a boiler. If waterworks are handy, connect the boiler with a hydrant and
after filling the boiler, let it receive the hydrant pressure.
Then examine the calked flues carefully, and if you see
any seeping of water, use your beader lightly till the water
stops. In case no waterworks with good pressure are at
hand, you can use a hydraulic pump or a good force
pump.

The amount of pressure required in testing a boiler
should be that at which the safety valve is set to blow off,
say 110 to 130 lbs. This will be sufficient.

If you are in the field with no hydrant or force pump
handy, you may test your boiler in this way: Take off
the safety valve and fill the boiler full of water through
the safety valve opening. Then screw the safety back in
its place. You should be sure that every bit of space in
the boiler is filled entirely full of water, with all openings
tightly closed. Then get back in the boiler and have a
bundle of straw burned under the firebox, or under the
waist of the boiler, so that at some point the water will be

slightly heated. This will cause pressure. If your safety valve is in perfect order, you will know as soon as water begins to escape at the safety valve whether your flues are calked tight enough or not.

The water is heated only a few degrees, and the pressure is cold water pressure. In very cold weather this method cannot be used, however, as water has no expansive force within five degrees of freezing.

The above methods are not intended for testing the safety of a boiler, but only for testing for leaky flues. If you wish to have your boiler tested, it is better to get an expert to do it.

CHAPTER V.

A traction engine is usually the simplest kind of an engine made. If it were not, it would require a highly expert engineer to run it, and this would be too costly for a farmer or thresherman contractor. Therefore the builders of traction engines make them of the fewest possible parts, and in the most durable and simple style. Still, even the simplest engine requires a certain amount of brains to manage it properly, especially if you are to get the maximum of work out of it at the lowest cost.

If the engine is in perfect order, about all you have to do is to see that all bearings are properly lubricated, and that the automatic oiler is in good working condition. But as soon as an engine has been used for a certain time, there will be wear, which will appear first in the journals, boxes and valve, and it is the first duty of a good engineer to adjust these. To adjust them accurately requires skill; and it is the possession of that skill that goes to make a real engineer.

Your first attention will probably be required for the cross-head and crank boxes or brasses. The crank box and pin will probably wear first; but both the cross-head and crank boxes are so nearly alike that what is said of one will apply to the other.

You will find the wrist box in two parts. In a new engine these parts do not quite meet. There is perhaps an eighth of an inch waste space between them. They are brought up to the box in most farm engines by a wedge-shaped key. This should be driven down a little at a time as the boxes wear, so as to keep them snug up to the pin, though not too tight.

You continue to drive in the key and tighten up the boxes as they wear until the two halves come tight together.   Then you can no longer accomplish anything in this way.

When the brasses have worn so that they can be forced no closer together, they must be taken off and the ends of them filed where they come together.   File off a sixteenth of an inch from each end.   Do it with care, and be sure you get the ends perfectly even.   When you have done this you will have another eighth of an inch to allow for wear.

Now, by reflection you will see that as the wrist box wears, and the wedge-shaped key is driven in, the pitman (or piston arm) is lengthened to the amount that the half of the box farthest from the piston has worn away.   When the brasses meet, this will amount to one-sixteenth of an inch.

Now if you file the ends off and the boxes wear so as to come together once more, the pitman will have been shortened one-eighth of an inch; and pretty soon the clearance of the piston in the cylinder will have been offset, and the engine will begin to pound.   In any case, the clearance at one end of the cylinder will be one-sixteenth or one-eighth of an inch less, and in the other end one-sixteenth or one-eighth of an inch more.   When this is the case you will find that the engine is not working well.

To correct this, when you file the brasses either of the cross-head box or the crank box you must put in some filling back of the brass farthest from the piston, sufficient to equalize the wear that has taken place, that is, one-sixteenth of an inch each time you have to file off a sixteenth of an inch.   This filling may be some flat pieces of tin or sheet copper, commonly called shims, and the process is called shimming.   As to the front half of the box, no shims are required, since the tapering key brings that box up to its proper place.

Great care must be exercised when driving in the tapering key or wedge to tighten up the boxes, not to drive it in too hard.   Many engineers think this is a sure remedy for "knocking" in an engine, and every time they hear a knock they drive in the crank box key.   Often the knock is from some other source, such as from a loose fly wheel, or the like. Your ear is likely to deceive you; for a knock from any part of an engine is likely to sound as if it came from the crank box. If you insist on driving in the key too hard and too often, you will ruin your engine.

In tightening up a key, first loosen the set screw that holds the key; then drive down the key till you think it is tight; then drive it back again, and this time force it down with your fist as far as you can. By using your fist in this way after you have once driven the pin in tight and loosened it again you may be pretty certain you are not going to get it so tight it will cause the box to heat.

WHAT CAUSES AN ENGINE TO KNOCK.

The most common sign that something is loose about an engine is "knocking," as it is called. If any box wears a little loose, or any wheel or the like gets a trifle loose, the engine will begin to knock.

When an engine begins to knock or run hard, it is the duty of the engineer to locate the knock definitely. He must not guess at it. When he has studied the problem out carefully, and knows where the knock is, then he may proceed to remedy it. Never adjust more than one part at a time.

As we have said, a knock is usually due to looseness somewhere. The journals of the main shaft may be loose and cause knocking. They are held in place by set bolts and jam nuts, and are tightened by simply screwing up the nuts. But a small turn of a nut may make the box so tight it will begin to heat at once. Great care should be taken in tightening up such a box to be sure not to get it too tight. Once a box begins to cut, it should be taken out and thoroughly cleaned.

Knocking may be due to a loose eccentric yoke. There is packing between the two halves of the yoke, and to tighten up you must take out a thin layer of this packing. But be careful not to take out too much, or the eccentric will stick and begin to slip.

Another cause of knocking is the piston rod loose in

the cross-head. If the piston rod is keyed to the cross-head it is less liable to get loose than if it were fastened by a nut; but if the key continues to get loose, it will be best to replace it with a new one.

Unless the piston rod is kept tight in the cross-head, there is liability of a bad crack. A small strain will bring the piston out of the cross-head entirely, when the chances are you will knock out one or both cylinder-heads. If a nut is used, there will be the same danger if it comes off. It should therefore be carefully watched. The best way is to train the ear to catch any usual sound, when loosening of the key or nut will be detected at once.

Another source of knocking is looseness of the cross-head in the guides. Provision is usually made for taking up the wear; but if there is not, you can take off the guides and file them or have them planed off. You should take care to see that they are kept even, so that they will wear smooth with the crosshead shoes.

If the fly-wheel is in the least loose it will also cause knocking, and it will puzzle you not a little to locate it. It may appear to be tight; but if the key is the least bit too narrow for the groove in the shaft, it will cause an engine to bump horribly, very much as too much "lead" will.

LEAD.

We have already explained what "lead" is. It is opening of the port at either end of the steam cylinder allowed by the valve when the engine is on a dead centre. To find out what the lead is, the cover of the steam chest must be taken off, and the engine placed at each dead centre in succession. If the lead is greater at one end than it is at the other, the valve must be adjusted to equalize it. As a rule the engine is adjusted with a suitable amount of lead if it is equalized. The correct amount of lead varies with the engine and with the port opening. If the port opening is long and narrow, the lead should obviously be less than if the port is short and wide.

If the lead is insufficient, there will not be enough steam let into the cylinder for cushion, and the engine will knock. If there is too much lead the speed of the engine will be lessened, and it will not do the work it ought. To adjust the lead *de novo* is by no means an easy task.

HOW TO SET A SIMPLE VALVE.

In order to set a valve the engine must be brought to a dead centre. This cannot be done accurately by the eye. An old engineer* gives the following directions for finding the dead centre accurately. Says he: "First provide yourself with a 'tram.' This is a rod of one-fourth inch iron about eighteen inches long, with two inches at one end bent over to a sharp angle. Sharpen both ends to a point. Fasten a block of hard wood somewhere near the face of the fly-wheel, so that when the straight end of your tram is placed at a definite point in the block, the hooked end will reach the crown of the fly-wheel. The block must be held firmly in its place, and the tram must always touch it at exactly the same point.

"You are now ready to set about finding the dead centre. In doing this, remember to turn the fly-wheel always in the same direction.

"Bring the engine over till it nearly reaches one of the dead centres, but not quite. Make a distinct mark across the cross-head and guides. Also go around to the fly-wheel, and placing the straight end of the tram at the selected point on the block of wood, make a mark across the crown or centre of face of the fly-wheel. Now turn your engine past the centre, and on to a point at which the mark on the cross head will once more exactly correspond with the line on the guides, making a single straight line. Once more place the tram as before and make another mark across the crown of the fly-wheel. By use of dividers, find the exact centre between the two marks made on the fly-wheel, and mark this point distinctly with a centre punch. Now bring the fly-wheel to the point where the tram, set with its straight end at the required point on the block of wood, will touch this point with the hooked end, and you will have one of the dead centres.

---

*J. H. Maggard.

"Turn the engine over and proceed in the same way to find the other dead centre."

Now, setting the engine on one of the dead centres, remove the cover of the steam chest and proceed to set your valve.

Assuming that the engine maker gave the valve the proper amount of lead in the first place, you can proceed on the theory that it is merely necessary to equalize the lead at both ends. Assume some convenient lead, as one-sixteenth of an inch, and set the valve to that. Then turn the engine over and see if the lead at the other end is the same. If it is the same, you have set the valve correctly. If it is less at the other end, you may conclude that the lead at both ends should be less than one-sixteenth of an inch, and must proceed to equalize it. This you can do by fitting into the open space a little wedge of wood, changing the valve a little until the wedge goes in to just the same distance at each end. Then you may know that the lead at one end is the same as at the other end. You can mark the wedge for forcing it against the metal, or mark it against the seat of the valve with a pencil.

The valve is set by loosening the set screws that hold the eccentric on the shaft. When these are loosened up the valve may be moved freely. When it is correctly set the screws should be tightened, and the relative position of the eccentric on the shaft may be permanently marked by setting a cold chisel so that it will cut into the shaft and the eccentric at the same time and giving it a smart blow with the hammer, so as to make a mark on both the eccentric and the shaft. Should your eccentric slip at any time in the future, you can set your valve by simply bringing the mark on the eccentric so that it will correspond with the mark on the shaft. Many engines have such a mark made when built, to facilitate setting a valve should the eccentric become loose.

These directions apply only to setting the valve of a single eccentric engine.

HOW TO SET A VALVE ON A DOUBLE ECCENTRIC ENGINE.

In setting a valve on a reversible or double eccentric engine, the link may cause confusion, and you may be trying to set the valve to run one way when the engine is set to run the other.

The valve on such an engine is exactly the same as on a single eccentric engine. Set the reverse lever for the engine to go forward. Then set the valve exactly as with a single eccentric engine. When you have done so, tighten the eccentric screws so that they will hold temporarily, and set the reverse lever for the engine to go backward. Then put the engine on dead centres and see if the valve is all right at both ends. If it is, you may assume that it is correctly set, and tighten eccentric screws, marking both eccentrics as before.

As we have said, most engines are marked in the factory, so that it is not a difficult matter to set the valves, it being necessary only to bring the eccentric around so that the mark on it will correspond with the mark on the shaft.

You can easily tell whether the lead is the same at both ends by listening to the exhaust. If it is longer at one end than the other, the valve is not properly set.

SLIPPING OF THE ECCENTRIC OR VALVE.

If the eccentric slips the least bit it may cause the engine to stop, or to act very queerly. Therefore the marks on the shaft and on the eccentric should be watched closely, and of course all grease and dirt should be kept wiped off, so that they can be seen easily. Then the jam nuts should be tightened up a little from time to time.

If the engine seems to act strangely, and yet the eccentrics are all right, look at the valve in the steam chest. If the valve stem has worked loose from the valve, trouble will be caused. It may be held in place by a nut, and the nut may work off; or the valve may be held by a clamp and pin, and the pin may work loose. Either will cause loss of motion, and perhaps a sudden stopping of the engine.

USE OF THE CYLINDER STEAM COCKS.

It is a comparatively simple matter to test a steam cylinder by use of the cylinder cocks. To do this, open both cocks, place the engine on the forward center, and turn on a little steam. If the steam blows out at the forward cock, we may judge that our lead is all right. Now turn the engine to the back center and let on the steam. It should blow out the same at the back cock. A little training of the ear will show whether the escape of steam is the same at both ends. Then reverse the engine, set it on each center successfully, and notice whether the steam blows out from one cock at a time and in the same degree of force.

If the steam blows out of both cocks at the same time, or out of one cock on one center, but not out of the other cock on its corresponding center, we may know something is wrong. The valve does not work properly.

We will first look at the eccentrics and see that they are all right. If they are, we must open the steam chest, first turning off all steam. Probably we shall find that the valve is loose on the valve rod, if our trouble was that the steam blew out of the cock but did not out of the other when the engine was on the opposite center.

If our trouble was that steam blew out of both cocks at the same time, we may conclude either that the cylinder rings leak or else the valve has cut its seat. It will be a little difficult to tell which at first sight. In any case it is a bad thing, for it means loss of power and waste of steam and fuel. To tell just where the trouble is you must take off the cylinder head, after setting the engine on the forward center. Let in a little steam from the throttle. If it blows through around the rings, the trouble is with them; but if it blows through the valve port, the trouble is with the valve and valve seat.

If the rings leak you must get a new set if they are of the self-adjusting type. But if they are of the spring or adjusting type you can set them out yourself; but few engines now use the latter kind of rings, so a new pair will probably be required.

If the trouble is in the valve and valve seat, you should take the valve out and have the seat planed down, and the valve fitted to the seat. This should always be done by a skilled mechanic fully equipped for such work, as a novice is almost sure to make bad work of it. The valve seat and valve must be scraped down by the use of a flat piece of very hard steel, an eighth of an inch thick and about 3 by 4 inches in size. The scraping edge must be absolutely straight. It will be a slow and tedious process, and a little too much scraping on one side or the other will prevent a perfect fit. Both valve and valve seat must be scraped equally. Novices sometimes try to reseat a valve by the use of emery. This is very dangerous and is sure to ruin the valve, as it works into the pores of the iron and causes cutting.

LUBRICATION.

A knowledge of the difference between good oil and poor oil, and of how to use oil and grease, is a prime essential for an engineer.

First let us give a little attention to the theory of lubrication. The oil or grease should form a lining between the journal and its pin or shaft. It is in the nature of a slight and frictionless cushion at all points where the two pieces of metal meet.

Now if oil is to keep its place between the bearing and the shaft or pin it must stick tight to both pieces of metal, and the tighter the better. If the oil is light the forces at work on the bearings will force the oil away and bring the metals together. As soon as they come together they begin to wear on each other, and sometimes the wear is very rapid. This is called "cutting." If a little sand or grit gets into the bearing, that will help the cutting wonderfully, and more especially if there is no grease there.

For instance, gasoline and kerosene are oils, but they are so light they will not stick to a journal, and so are valueless for lubricating. Good lubricating oil will cost a little more than cheap oil which has been mixed with worthless oils to increase its bulk without increasing its cost. The higher priced oil will really cost less in the end, because there is a larger percentage of it which will do service. A good engineer will have it in his contract that he is to be furnished with good oil.

Now an engine requires two different kinds of oil, one for the bearings, such as the crank pin, the cross-head and journals, and quite a different kind for lubricating the steam cylinder.

It is extremely important that the steam cylinder should be well lubricated; and this cannot be done direct. The oil must be carried into the valve and cylinder with steam. The heat of the steam, moreover, ranging from about 320 degrees Fahr. for 90 lbs. pressure to 350 degrees for 125 lbs. of pressure, will quickly destroy the efficacy of a poor oil, and a good cylinder oil must be one that will stick to the cylinder and valve seat under this high temperature. It must have staying qualities.

The link reverse is one of the best for its purpose; but it requires a good quality of oil on the valve for it to work well. If the valve gets a little dry, or the poor oil used does not serve its purpose properly, the link will begin to jump and pound. This is a reason why makers are substituting other kinds of reverse gear in many ways not as good, but not open to this objection. If a link reverse begins to pound when you are using good oil, and the oiler is working properly, you may be sure something is the matter with the valve or the gear.

A good engineer will train his ear so that he will detect by simply listening at the cylinder whether everything is working exactly as it ought. For example, the exhaust at each end of the cylinder, which you can hear distinctly, should be the same and equal. If the exhaust at one end is less than it is at the other, you may know that one end of the cylinder is doing more work than the other. And also any little looseness or lack of oil will signify itself by the peculiar sound it will cause.

While the cylinder requires cylinder oil, the crank, cross-head and journals require engine oil, or hard grease. The use of hard grease is rapidly increasing, and it is highly to be recommended. With a good automatic spring grease cup hard grease will be far less likely to let the bearings heat than common oil will. At the same time it will be much easier to keep an engine clean if hard grease is used.

An old engineer* gives the following directions for fitting a grease cup on a box not previously arranged for one: "Remove the journal, take a gouge and cut a clean groove across the box, starting at one corner, about one-eighth of an inch from the point of the box, and cut diagonally across, coming out at the opposite corner on the other end of the box. Then start at the opposite corner and run through as before, crossing the first groove in the center of the box. Groove both halves of the box the same, being careful not to cut out at either end, as this will allow the grease to escape from the box and cause unnecessary waste. The shimming or packing in the box should be cut so as to touch the journal at both ends of the box, but not in the center or between these two points. So when the top box is brought down tight this will form another reservoir for the grease. If the box is not tapped directly in the center for the cup, it will be necessary to cut another groove from where it is tapped into the grooves already made. A box prepared in this way and carefully polished inside, will require little attention if you use good grease."

A HOT BOX.

When a box heats in the least degree, it is a sign that for lack of oil or for some other reason the metals are wearing together.

The first thing to do, of course, is to see that the box is supplied with plenty of good oil or grease.

If this does not cause the box to cool off, take it apart and clean it thoroughly. Then coat the journal with white lead mixed with good oil. Great care should be exercised to keep all dirt or grit out of your can of lead and away from the bearing.

Replace the oil or grease cup, and the box will soon cool down.

*J. H. Maggard.

THE FRICTION CLUTCH.

Nearly all traction engines are now provided with the friction clutch for engaging the engine with the propelling gear. The clutch is usually provided with wooden shoes,

A. W. STEVENS CO. FRICTION CLUTCH.

which are adjustable as they wear; and the clutch is thrown on by a lever, conveniently placed.

Before running an engine, you must make sure that the clutch shoes are properly adjusted. Great care must be taken to be sure that both shoes will come in contact with the friction wheel at the same instant; for if one shoe touches the wheel before the other the clutch will probably slip.

The shoes should be so set as to make it a trifle difficult to draw the lever clear back.

To regulate the shoes on the Rumely engine, for example, first throw the friction in. The nut on the top of the toggle connecting the sleeve of the friction with the shoe must then be loosened, and the nut below the shoe tightened up, forcing the shoe toward the wheel. Both shoes should be carefully adjusted so that they will engage the band wheel equally and at exactly the same time.

To use the friction clutch, first start the engine, throwing the throttle gradually wide open. When the engine is running at its usual speed, slowly bring up the clutch until the gearing is fully engaged, letting the engine start slowly and smoothly, without any jar.

Traction engines having the friction clutch are also provided with a pin for securing a rigid connection, to be used in cases of necessity, as when the clutch gets broken or something about it gives out, or you have difficulty in making it hold when climbing hills. This pin is a simple round or square pin that can be placed through a hole in one of the spokes of the band wheel until it comes into a similar opening in the friction wheel. When the pin is taken out, so as to disconnect the wheels, it must be entirely removed, not left sticking in the hole, as it is liable to catch in some other part of the machinery.

MISCELLANEOUS SUGGESTIONS.

Be careful not to open the throttle valve too quickly, or you may throw off the driving belt. You may also stir up the water and cause it to pass over with the steam, starting what is called "priming."

Always open your cylinder cocks when you stop, to

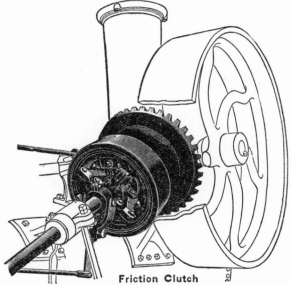

Friction Clutch

AULTMAN & TAYLOR FRICTION CLUTCH.

make sure all water has been drained out of the cylinder; and see that they are open when you start, of course closing them as soon as the steam is let in.

When you pull out the ashes always have a pail of water ready, for you may start a fire that will do no end of damage.

If the water in your boiler gets low and you are waiting for the tank to come up, don't think you "can keep on a little longer," but stop your engine at once. It is better to lose a little time than run the risk of an explosion that will ruin your reputation as an engineer and cause your employer a heavy expense.

Never start the pump when the water in the boiler is low.

Be sure the exhaust nozzle does not get limed up, and be sure the pipe where the water enters the boiler from the heater is not limed up, or you may split a heater pipe or knock out a check valve.

Never leave your engine in cold weather without draining off all the water; and always cover up your engine when you leave it.

Never disconnect the engine with a leaky throttle.

Keep the steam pressure steady, not varying more than 10 to 15 lbs.

If called on to run an old boiler, have it thoroughly tested before you touch it.

Always close your damper before pulling through a stack yard.

Examine every bridge before you pull on to it.

Do not stop going down a steep grade.

CHAPTER VI.

It is something of a trick to handle a traction engine on the road. The novice is almost certain to run it into a ditch the first thing, or get stuck on a hill, or in a sand patch or a mudhole. Some attention must therefore be paid to handling a traction engine on the road.

In the first place, never pull the throttle open with a jerk, nor put down the reverse lever with a snap. Handle your engine deliberately and thoughtfully, knowing beforehand just what you wish to do and how you will do it. A traction engine is much like an ox; try to goad it on too fast and it will stop and turn around on you. It does its best work when moving slowly and steadily, and seldom is anything gained by rushing.

The first thing for an engineer to learn is to handle his throttle. When an engine is doing work the throttle should be wide open; but on the road, or in turning, backing, etc., the engineer's hand must be on the throttle all the time and he must exercise a nice judgment as to just how much steam the engine will need to do a certain amount of work. This the novice will find out best by opening the throttle slowly, taking all the time he needs, and never allowing any one to hurry him.

As an engineer learns the throttle, he gradually comes to have confidence in it. As it were, he feels the pulse of the animal and never makes a mistake. Such an engineer always has power to spare, and never wastes any power. He finds that a little is often much better than too much.

The next thing to learn is the steering wheel. It has tricks of its own, which one must learn by practice. Most young engineers turn the wheel altogether too much. If you let your engine run slowly you will have time to turn the wheel slowly, and accomplish just what you want to do. If you hurry you will probably have to do your work

all over again, and so lose much more time in the end
than if you didn't hurry.

Always keep your eyes on the front wheels of the engine, and do not turn around to see how your load is coming on. Your load will take care of itself if you manage
the front wheels all right, for they determine where you
are to go.

In making a hard turn, especially, go slow. Then you
will run no chance of losing control of your engine, and
you can see that neither you nor your load gets into a
ditch.

GETTING INTO A HOLE.

You are sure sooner or later to get into a hole in the
road, for a traction engine is so heavy it is sure to find
any soft spot in the road there may be.

As to getting out of a hole, observe in the first place
that you must use your best judgment.

First, never let the drive wheels turn round without
doing any work. The more they spin round without
helping you, the worse it will be for you.

Your first thought must be to give the drive wheels
something they can climb on, something they can stick
to. A heavy chain is perhaps the very best thing you can
put under them. But usually on the road you have no
chain handy. In that case, you must do what you can. Old
hay or straw will help you; and so will old rails or any
old timber.

Spend your time trying to give your wheels something
to hold to, rather than trying to pull out. When the
wheels are all right, the engine will go on its way without any trouble whatever. And do not half do your
work of fixing the wheels before you try to start. See
that both wheels are secure before you put on a pound of
steam. Make sure of this the first time you try, and you
will save time in the end. If you fix one wheel and don't
fix the other, you will probably spoil the first wheel by
starting before the other is ready.

Should you be where your engine will not turn, then
you are stuck indeed. You must lighten your load or dig
a way out.

BAD BRIDGES.

A traction engine is so heavy that the greatest care must be exercised in crossing bridges. If a bridge floor is worn, if you see rotten planks in it, or liability of holes, don't pull on to that bridge without taking precautions.

The best precaution is to carry with you a couple of planks sixteen feet long, three inches thick in the middle, tapering to two inches at the ends ; also a couple of planks eight feet long and two inches thick, the latter for culverts and to help out on long bridges.

Before pulling on to a bad looking bridge, lay down your planks, one for each pair of wheels of the engine to run on. Be exceedingly careful not to let the engine drop off the edge of these planks on the way over, or pass over the ends on to the floor of the bridge. If one pair of planks is too short, use your second pair.

Another precaution which it is wise to take is to carry fifty feet of good, stout hemp rope, and when you come to a shaky bridge, attach your separator to the engine by this rope at full length, so that the engine will have crossed the bridge before the weight of the separator comes upon it.

Cross a bad bridge very slowly. Nothing will be gained by hurrying. There should especially be no sudden jerks or starts.

SAND PATCHES.

A sandy road is an exceedingly hard road to pull a load over.

In the first place, don't hurry over sand. If you do you are liable to break the footing of the wheels, and then you are gone.

In the second place, keep your engine as steady and straight as possible, so that both wheels will always have an equal and even bearing. They are less liable to slip if you do. It is useless to try to "wiggle" over a sand patch. Slow, steady, and even is the rule.

If your wheels slip in sand, a bundle of straw or hay,

especially old hay, will be about the best thing to give them a footing.

In climbing hills take the same advice we have given you all along: Go slow. Nothing is gained by rushing at a hill with a steam engine. Such an engine works best when its force is applied steadily and evenly, a little at a time.

If you have a friction clutch, as you probably will have, you should be sure it is in good working order before you attempt to climb hills. It should be adjusted to a nicety, as we have already explained. When you come to a bad hill it would probably be well to put in the tight gear pin; or use it altogether in a hilly country.

When the friction clutch first came into use, salesmen and others used to make the following recommendation (a recommendation which we will say right here is bad). They said, when you come to an obstacle in the road that you can't very well get your engine over, throw off your friction clutch from the road wheels, let your engine get under good headway running free, and then suddenly put on the friction clutch and jerk yourself over the obstacle.

Now this is no doubt one way to get over an obstacle; but no good engineer would take his chances of spoiling his engine by doing any such thing with it. Some part of it would be badly strained by such a procedure; and if this were done regularly all through a season, an engine would be worth very little at the end of the season.

CHAPTER VII.

QUESTIONS AND ANSWERS.

THE BOILER.

Q. How should water be fed to a boiler?

A. In a steady stream, by use of a pump or injector working continuously and supplying just the amount of water required. By this means the water in the boiler is maintained at a uniform level, and produces steam most evenly and perfectly.

Q. Why should pure water be used in a boiler ?

A. Because impure water, or hard water, forms scales on the boiler flues and plates, and these scales act as non-conductors of heat. Thus the heat of the furnace is not able to pass easily through the boiler flues and plates to the water, and your boiler becomes what is called "a hard steamer."

Q. What must be done to prevent the formation of scale?

A. First, use some compound that will either prevent scale from forming, or will precipitate the scale forming substance as a soft powder that can easily be washed off. Sal soda dissolved in the feed water is recommended, but great care should be exercised in the use of sal soda not to use too much at a time, as it may cause a boiler to foam. Besides using a compound, clean your boiler often and regularly with a hand hose and a force pump, and soak it out as often as possible by using rain water for a day or two, especially before cleaning. Rain water will soften and bring down the hard scale far better than any compound.

Q. How often should you clean your boiler?

A. As often as it needs it, which will depend upon the work you do and the condition of the water. Once a

week is usually often enough if the boiler is blown down
a little every day.   If your water is fairly good, once a
month will be often enough.   A boiler should be blown
off about one gauge at a time two or three times a day
with the blow-off if the water is muddy.

Q.   How long should the surface blow-off be left
open?

A.   Only for a few seconds, and seldom longer than
a minute.   The surface blow-off carries off the scum that
forms on the water, and other impurities that rise with
the scum.

Q.   How do you clean a boiler by blowing off?

A.   When the pressure has been allowed to run down
open the blow-off valve at the bottom of the boiler and let
the water blow out less than a minute, till the water drops
out of sight in the water gauges, or about two and one-
half inches.   Blown off more is only a waste of heat and
fuel.

Q.   What harm will be done by blowing off a boiler
under a high pressure of steam?

A.   The heat in the boiler while there is such a pressure will be so great that it will bake the scale on the
inside of the boiler, and it will be very difficult to remove
it afterward. After a boiler has been blown off the scale
should be for the most part soft, so that it can be washed
out by a hose and force pump.

Q.   Why should a hot boiler never be filled with cold
water?

A.   Because the cold water will cause the boiler to
contract more in some places than in others, and so suddenly that the whole will be badly strained.   Leaky flues
are made in this way, and the life of a boiler greatly
shortened.   As a rule a boiler should be filled only when
the metal and the water put into it are about at the same
temperature.

Q.   After a boiler has been cleaned, how should the
manhole and manhole plates be replaced?

A.   They are held in position by a bolt passing through
a yoke that straddles the hole; but to be steam and water
tight they must have packing all around the junction of
the plate with the boiler.   The best packing is sheet rub-

ber cut in the form of a ring just the right size for the bearing surface. Hemp or cotton packing are also used, but they should be free from all lumps and soaked in oil. Do not use any more than is absolutely needed. Be careful, also, to see that the bearings of the plate and boiler are clean and smooth, with all the old packing scraped off. Candle wick saturated with red lead is next best to rubber as packing.

Q. What are the chief duties of an engineer in care of a boiler?

A. First, to watch all gauges, fittings, and working parts, to see that they are in order; try the gauge cocks to make sure the water is at the right height; try the safety valve from time to time to be sure it is working; see that there are no leaks, that there is no rusting or wearing of parts, or to replace parts when they do begin to show wear; to examine the check valve frequently to make sure no water can escape through it from the boiler; take precautions against scale and stoppage of pipes by scale; and keep the fire going uniformly, cleanly, and in an economical fashion.

Q. What should you do if the glass water gauge breaks?

A. Turn off the gauge cocks above and below, the lower one first so that the hot water will not burn you. You may put in a new glass and turn on gauge cocks at once. Turn on the lower or water cock first, then the upper or steam cock. You may go on without the glass gauge, however, using the gauge cocks or try cocks every few minutes to make sure the water is at the right height, neither too high nor too low.

Q. Why is it necessary to use the gauge cocks when the glass gauge is all right?

A. First, because you cannot otherwise be sure that the glass gauge is all right; and, secondly, because if you do not use them frequently they are likely to become scaled up so that you cannot use them in case of accident to the glass gauge.

Q. If a gauge cock gets leaky, what should be done?

A. Nothing until the boiler has cooled down. Then if the leak is in the seat, take it out and grind and refit it;

if the leak is where the cock is screwed into the boiler, tighten it up another turn and see if that remedies the difficulty. If it does not you will probably have to get a new gauge cock.

Q. Why not screw up a gauge cock while there is a pressure of steam on?

A. The cock might blow out and cause serious injury to yourself or some one else. Make it a rule never to fool with any boiler fittings while there is a pressure of steam on the boiler. It is exceedingly dangerous.

Sometimes a gauge cock gets broken off accidentally while the boiler is in use. If such an accident happens, bank the fire by closing the draft and covering the fire with fresh fuel or ashes. Stop the engine and let the water blow out of the hole till only steam appears; then try to plug the opening with a long whitewood or poplar, or even a pine stick (six or eight feet long), one end of which you have whittled down to about the size of the hole. When the steam has been stopped the stick may be cut off close to the boiler and the plug driven in tight. If necessary you may continue to use the boiler in this condition until a new cock can be put in.

Q. What should you do when a gauge cock is stopped up?

A. Let the steam pressure go down, and then take off the front part and run a small wire into the passage, working the wire back and forth until all scale and sediment has been removed.

Q. What should you do when the steam gauge gets out of order.

A. If the steam gauge does not work correctly, or you suspect it does not, you may test it by running the steam up until it blows off at the safety valve. If the steam gauge does not indicate the pressure at which the safety valve is set to pop off, and you have reason to suppose the safety valve is all right, you may conclude that there is something the matter with the steam gauge. In that case either put in a new one, or, if you have no extra steam gauge on hand, shut down your boiler and engine till you can get your steam gauge repaired. Sometimes this can be done simply by adjusting the pointer, which may have got loose, and you can test it by attaching it to another boiler which has a steam gauge that is all right and by which you can check up yours. It is VERY DANGEROUS to run your boiler without a steam gauge, depending on the safety valve. Never allow the slightest variation in correctness of the steam gauge without repairing it at once. It will nearly always be cheaper in these days to put in a new gauge rather than try to repair the old one.

Q. What should you do if the pump fails to work?

A. Use the injector.

Q. What should you do if there is no injector?

A. Stop the engine at once and bank the fire with damp ashes, especially noting that the water does not fall below the bottom of the glass gauge. Then examine the pump. First see if the plunger leaks air; if it is all right, examine the check valves, using the little drain cock as previously explained to test the upper ones, for the valves may have become worn and will leak; third, if the check valves are all right, examine the supply pipe, looking at the strainer, observing whether suction takes place when the pump is worked, etc. There may be a leak in the suction hose somewhere during its course where air can get in, or it may become weak and collapse under the force of the atmosphere, or the lining of the suction pipe may have become torn or loose. The slightest leak in the suction pipe will spoil the working of the pump. Old tubing should never be used, as it is sure to give trouble. Finally, examine the delivery pipe. Close the cock or valve next the boiler, and examine the boiler check valve; notice whether the pipe is getting limed up. If necessary, disconnect the pipe and clean it out with a stiff wire. If everything is all right up to this point, you must let the boiler cool off, blow out the water, disconnect the pipe between the check and the boiler, and thoroughly clean the delivery pipe into the boiler. Stoppage of the delivery pipe is due to deposits of lime from the heating of the water in the heater. Stoppage from this source will be gradual, and you will find less and less water going into your boiler from your pump until none flows at all. From this you may guess the trouble.

Q. How may the communication with the water gauge always be kept free from lime?

A. By blowing it off through the drain cock at the bottom. First close the upper cock and blow off for a few seconds, the water passing through the lower cock; then close the lower cock and open the upper one, allowing the steam to blow through this and the drain cock for a few seconds. If you do this every day or oftener you will have no trouble.

Q. Should the water get low for any reason, what should be done?

A. Close all dampers tight so as to prevent all draft, and bank the fire with fresh fuel or with ashes (damp ashes are the best if danger is great). Then let the boiler cool down before putting in fresh water. Banking the fire is better than drawing or dumping it, as either of these make the heat greater for a moment or two, and that additional heat might cause an explosion. Dashing cold water upon the fire is also very dangerous and in every way unwise. Again, do not open the safety valve, for that also, by relieving some of the pressure on the superheated water, might cause it to burst suddenly into steam and so cause an explosion.

Q. Under such circumstances, would you stop the engine?

A. No; for a sudden checking of the outflow of steam might bring about an explosion. Do nothing but check the heat as quickly and effectively as you can by banking or covering the fires.

Q. Why not turn on the feed water?

A. Because the crown sheet of the boiler has become overheated, and any cold water coming upon it would cause an explosion. If the pump or injector are running, of course you may let them run, and the boiler will gradually refill as the heat decreases. Under such circumstances low water is due to overheating the boiler.

Q. Would not the fusible plug avert any disaster from low water?

A. It might, and it might not. The top of it is liable to get coated with lime so that the device is worthless. You should act at all times precisely as if there were no fusible plug. If it ever does avert an explosion you may be thankful, but averting explosions by taking such means as we have suggested will be far better for an engineer's reputation.

Q. Would not the safety valve be a safeguard against explosion?

A. No; only under certain conditions. It prevents too high a pressure for accumulating in the boiler when there is plenty of water; but when the water gets low the safety valve may only hasten the explosion by relieving some of the pressure and allowing superheated water to burst suddenly into steam, thus vastly expanding instantly.

Q. Should water be allowed to stand in the boiler when it is not in use?

A. It is better to draw it off and clean the boiler, to prevent rusting, formation of scale, hardening of sediment, etc., if boiler is to be left for any great length of time.

Q. What should you do if a grate bar breaks or falls out?

A. You should always have a spare grate bar on hand to put in its place; but if you have none you may fill the space by wedging in a stick of hard wood cut the right shape to fill the opening. Cover this wood with ashes before poking the fire over it, and it will last for several hours before it burns out. You will find it exceedingly difficult to keep up the fire with a big hole in the grate that will let cold air into the furnace and allow coal to drop down.

In case the grate is of the rocker type the opening may be filled by shaping a piece of flat iron, which can be set in without interfering with the rocking of the grate; or the opening may be filled with wood as before if the wood is covered well with ashes. Of course the use of wood will prevent the grate from rocking and the poker must be used to clean.

Q. Why should an engineer never start a boiler with a hot fire, and never let his fire get hotter than is needed to keep up steam?

A. Both will cause the sheets to warp and the flues to become leaky, because under high heat some parts of the boiler will expand more rapidly than others. For a similar reason, any sudden application of cold to a boiler, either cold water or cold air through the firebox door, will cause quicker contraction of certain parts than other parts, and this will ruin a boiler.

Q. How should you supply a boiler with water?

A. In a regular stream continually. Only by making the water pass regularly and gradually through the heater will you get the full effect of the heat from the exhaust steam. If a great deal of water is pumped into the boiler at one time, the exhaust steam will not be sufficient to heat it as it ought. Then if you have a full boiler and shut off the water supply, the exhaust steam in the heater is wasted, for it can do no work at all. Besides, it hurts the boiler to allow the temperature to change, as it will inevitably do if water is supplied irregularly.

WHATEVER YOU DO, NEVER ATTEMPT TO TIGHTEN A SCREW OR CALK A BOILER UNDER STEAM PRESSURE. IF ANYTHING IS LOOSE IT IS LIABLE TO BLOW OUT IN YOUR FACE WITH DISASTROUS CONSEQUENCES.

Q. If boiler flues become leaky, can an ordinary person tighten them?

A. Yes, if the work is done carefully. See full explanation previously given, p. 17. Great care should be taken not to expand the flues too much, for by so doing you are likely to loosen other flues and cause more leaks than you had in the first place. Small leaks inside a boiler are not particularly dangerous, but they should be remedied at the earliest possible moment, since they reduce the power of the boiler and put out the fire. Besides, they look bad for the engineer.

Q. How should flues be cleaned?

A. Some use a steam blower; but a better way is to scrape off the metal with one of the many patent scrapers, which just fill the flue, and when attached to a rod and worked back and forth a few times the whole length of the flue do admirable service.

Q. What harm will dirty flues do?

A. Two difficulties arise from dirty flues. If they become reduced in size the fire will not burn well. Then, the same amount of heat will do far less work because it is so much harder for it to get through the layer of soot and ashes, which are non-conductors.

Q. What would you do if the throttle broke?

A. Use reverse lever.

CHAPTER VIII.

QUESTIONS AND ANSWERS.

THE ENGINE.

Q. What is the first thing to do with a new engine?

A. With some cotton waste or a soft rag saturated with benzine or turpentine clean off all the bright work; then clean every bearing, box and oil hole, using a force pump with air current first, if you have a pump, and then wiping the inside out clean with an oily rag, using a wire if necessary to make the work thorough. If you do not clean the working parts of the engine thus before setting it up, grit will get into the bearings and cause them to cut. Parts that have been put together need not be taken apart; but you should clean everything you can get at, especially the oil holes and other places that may receive dirt during transportation.

After the oil holes have been well cleaned, the oil cups may be wiped off and put in place, screwing them in with a wrench.

Q. What kind of oil should you use?

A. Cylinder oil only for the cylinder; lard oil for the bearings, and hard grease if your engine is provided with hard grease cup for the cross-head and crank. The only good substitute for cylinder oil is pure beef suet tried out. Merchantable tallow should never be used, as it contains acid.

Q. Can fittings be screwed on by hand only?

A. No; all fittings should be screwed up tight with a wrench.

Q. When all fittings are in place, what must be done before the engine can be started?

A. See that the grates in the firebox are in place and all right; then fill the boiler with clean water until it

shows an inch to an inch and a half in the water gauge.
Start your fire, and let it burn slowly until there is a
pressure in the boiler of 10 or 15 lbs. Then you can turn
on the blower to get up draft. In the meantime fill all
the oil cups with oil; put grease on the gears; open and
close all cocks to see that they work all right; turn your
engine over a few times to see that it works all right;
let a little steam into the cylinder with both cylinder
cocks open—just enough to show at the cocks without
moving the engine—and slowly turn the engine over,
stopping it on the dead centers to see if the steam comes
from only one of the cylinder cocks at a time, and that
the proper one; reverse the engine and make the same
test. Also see that the cylinder oiler is in place and ready
for operation. See that the pump is all right and in
place, with the valve in the feedpipe open and also the
valve in the supply pipe.

By going over the engine in this way you will notice
whether everything is tight and in working order, and
whether you have failed to notice any part which you do
not understand. If there is any part or fitting you do
not understand, know all about it before you go ahead.

Having started your fire with dry wood, add fuel gradually, a little at a time, until you have a fire covering
every part of the grate. Regulate the fire by the damper
alone, never opening the firebox door even if the fire
gets too hot.

Q. In what way should the engine be started?

A. When you have from 25 to 40 lbs. of pressure
open the throttle valve a little, allowing the cylinder
cocks to be open also. Some steam will condense at first
in the cold cylinder, and this water must be allowed to
drain off. See that the crank is not on a dead center,
and put on just enough steam to start the engine. As
soon as it gets warmed up, and only dry steam appears
at the cocks, close the cylinder cocks, open the throttle
gradually till it is wide open, and wait for the engine to
work up to its full speed.

Q. How is the speed of the engine regulated??

A. By the governor, which is operated by a belt running to the main shaft. The governor is a delicate apparatus, and should be watched closely. It should move up and down freely on the stem, which should not leak steam. If it doesn't work steadily, you should stop the engine and adjust it, after watching it for a minute or two to see just where the difficulty lies.

Q. Are you likely to have any hot boxes?

A. There should be none if the bearings are all clean and well supplied with oil. However, in starting a new engine you should stop now and then and examine every bearing by laying your hand upon it. Remember the eccentric, the link pin, the cross-head, the crank pin. If there is any heat, loosen the boxes up a trifle, but only a very little at a time. If you notice any knocking or pounding, you have loosened too much, and should tighten again.

Q. What must you do in regard to water supply?

A. After the engine is started and you know it is all right, fill the tank on the engine and start the injector. It may take some patience to get the injector started, and you should carefully follow the directions previously given and those which apply especially to the type of injector used. Especially be sure that the cocks admitting the water through the feed pipe and into the boiler are open.

Q. Why are both a pump and an injector required on an engine?

A. The pump is most economical, because it permits the heat in the exhaust steam to be used to heat the feed water, while the injector heats the water by live steam. There should also be an injector, however, for use when the engine is not working, in order that the water in the boiler may be kept up with heated water. If a cross-head pump is used, of course, it will not operate when the engine is not running; and in case of an independent pump the heater will not heat the water when the engine is not running because there is little or no exhaust steam available. There is an independent pump (the Marsh pump) which heats the water before it goes into the boiler, and this may be used when the engine is shut down instead of the injector.

Q. What is the next thing to test?

A. The reversing mechanism. Throw the reverse lever back, and see if the engine will run equally well in the opposite direction. Repeat this a few times to make sure that the reverse is in good order.

Q. How is a traction engine set going upon the road?

A. Most traction engines now have the friction clutch. When the engine is going at full speed, take hold of the clutch lever and slowly bring the clutch against the band wheel. It will slip a little at first, gradually engaging the gears and moving the outfit. Hold the clutch lever in one hand, while with the other you operate the steering wheel. By keeping your hand on the clutch lever you may stop forward motion instantly if anything goes wrong. When the engine is once upon the road, the clutch lever may set in the notch provided for it, and the engine will go at full speed. You can then give your entire attention to steering.

Q. What should you do if the engine has no friction clutch?

A. Stop the engine, placing the reversing lever in the center notch. Then slide the spur pinion into the gear and open the throttle valve wide. You are now ready to control the engine by the reversing lever. Throw the lever forward a little, bringing it back, and so continue until you have got the engine started gradually. When well under way throw the reverse lever into the last notch, and give your attention to steering.

Q. How should you steer a traction engine?

A. In all cases the same man should handle the throttle and steer the engine. Skill in steering comes by practice, and about the only rule that can be given is to go slow, and under no circumstances jerk your engine about. Good steering depends a great deal on natural ability to judge distances by the eye and power by the feel. A good engineer must have a good eye, a good ear, and a good touch (if we may so speak). If either is wanting, success will be uncertain.

Q. How should an engine be handled on the road?

A. There will be no special difficulty in handling an engine on a straight, level piece of road, especially if the road is hard and without holes. But when you come to your first hill your troubles will begin.

Before ascending a hill, see that the water in the boiler does not stand more than two inches in the glass gauge. If there is too much water, as it is thrown to one end of the engine by the grade it is liable to get into the steam cylinder. If you have too much water, blow off a little from the bottom blow-off cock.

In descending a hill never stop your engine for a moment, since your crown sheet will be uncovered by reason of the water being thrown forward, and any cessation in the jolting of the engine which keeps the water flowing over the crown sheet will cause the fusible plug to blow out, making delay and expense.

Make it a point never to stop your engine except on the level.

Before descending a hill, shut off the steam at the throttle, and control the engine by the friction brake ; or if there is no brake, do not quite close the throttle, but set the reverse lever in the center notch, or back far enough to control the speed. It is seldom necessary to use steam in going down hill, however, and if the throttle is closed even with no friction brake, the reverse may be used in such a way as to form an air brake in the cylinder.

Get down to the bottom of a hill as quickly as you can.

Before descending a hill it would be well to close your dampers and keep the firebox door closed tight all the time. Cover the fire with fresh fuel so as to keep the heat down.

The pump or injector must be kept at work, however, since as you have let the water down low, you must not let it fall any lower or you are likely to have trouble.

In ascending a hill, do just the reverse, namely : Keep your fire brisk and hot, with steam pressure ascending ; and throw the reverse lever in the last notch, giving the engine all the steam you can, else you may get stuck. If you stop you are likely to overheat forward end of fire tubes. You are less liable to get stuck if you go slowly than if you go fast. Regulate speed by friction clutch.

CHAPTER IX.

MISCELLANEOUS.

Q. What is Foaming?

A. The word is used to describe the rising of water in large bubbles or foam. You will detect it by noticing that the water in the glass gauge rises and falls, or is foamy. It is due to sediment in the boiler, or grease and other impurities in the feed supply. Shaking up the boiler will start foaming sometimes; at other times it will start without apparent cause. In such cases it is due to the steam trying to get through a thick crust on the surface of the water.

Q. How may you prevent foaming?

A. It may be checked for a moment by turning off the throttle, so giving the water a chance to settle. It is generally prevented by frequently using the surface blow-off to clear away the scum. Of course the water must be kept as pure as possible, and especially should alkali water be avoided.

Q. What is priming?

A. Priming is not the same as foaming, though it is often caused by foaming. Priming is the carrying of water into the steam cylinder with the steam. It is caused by various things beside foaming, for it may be found when the boiler is quite clean. A sudden and very hot fire may start priming. Priming sometimes follows lowering of the steam pressure. Often it is due to lack of capacity in the boiler, especially lack of steam space, or lack of good circulation.

Q. How can you detect priming?

A. By the clicking sound it makes in the steam cylinder. The water in the gauge will also go up and down violently. There will also be a shower of water from the exhaust.

Q. What is the proper remedy for priming?

A. If it is due to lack of capacity in the boiler nothing can be done but get a new boiler. In other cases it may be remedied by carrying less water in the boiler when that can be done safely, by taking steam from a different point in the steam dome, or if there is no dome by using a long dry pipe with perforation at the end.

A larger steam pipe may help it; or it may be remedied by taking out the top row of flues.

Leaky cylinder rings or a leaky valve may also have something to do with it. In all cases these should be made steam tight. If the exhaust nozzle is choked up with grease or sediment, clean it out.

A traction engine with small steam ports would prime quickly under forced speed.

Q. How would you bank your fires?

A. Push the fire as far to the back of the firebox as possible and cover it over with very fine coal or with dry ashes. As large a portion as possible of the grate should be left open, so that the air may pass over the fire. Close the damper tight. By banking your fires at night you keep the boiler warm and can get up steam more quickly in the morning.

Q. When water is left in the boiler with banked fire in cold weather, what precautions ought to be taken?

A. The cocks in the glass water gauge should be closed and the drain cock at the bottom opened, for fear the water in the exposed gauge should freeze. Likewise all drain cocks in steam cylinder and pump should be opened.

Q. How should a traction engine be prepared for laying up during the winter?

A. First, the outside of the boiler and engine should be thoroughly cleaned, seeing that all gummy oil or grease is removed. Then give the outside of the boiler and smokestack a coat of asphalt paint, or a coat of lampblack and linseed oil, or at any rate a doping of grease.

The outside of the boiler should be cleaned while it is hot, so that grease, etc., may be easily removed while soft.

After the outside has been attended to, blow out the water at low pressure and thoroughly clean the inside in the usual way, taking out the handhole and manhole plates, and scraping off all scale and sediment.

After the boiler has been cleaned on the inside, fill it nearly full of water, and pour upon the top a bucket of black oil. Then let the water out through the blow-off at the bottom. As the water goes down it will have a coating of oil down the sides of the boiler.

All the brass fittings should be removed, including gauge cocks, check valves, safety valve, etc. Disconnect all pipes that may contain water, to be sure none remains in any of them. Open all stuffing boxes and take out packing, for the packing will cause the parts they surround to rust.

Finally, clean out the inside of the firebox and the fire flues, and give the ash-pan a good coat of paint all over, inside as well as out.

The inside of the cylinder should be well greased, which can be done by removing the cylinder head.

See that the top of the smoke stack is covered to keep out the weather.

All brass fittings should be carefully packed and put away in a dry place.

A little attention to the engine when you put it up will save twice as much time when you take it out next season, and besides save many dollars of value in the life of the engine.

Q. How should belting be cared for?

A. First, keep belts free from dust and dirt.

Never overload belts.

Do not let oil or grease drip upon them.

Never put any sticky or pasty grease on a belt.

Never allow any animal oil or grease to touch a rubber belt, since it will destroy the life of the rubber.

The grain or hair side should run next the pulley, as it holds better and is not so likely to slip.

Rubber belts will be greatly improved if they are covered with a mixture of black lead and litharge, equal parts, mixed with boiled oil, and just enough japan to dry them quickly. This mixture will do to put on places that peel.

Q. What is the proper way to lace a belt?

A. First, square the ends with a proper square, cutting them off to a nicety. Begin to lace in the middle, and do not cross the laces on the pulley side. On that side the lacings should run straight with the length of the belt.

The holes in the belt should be punched if possible with an oval punch, the long diameter coinciding with the length of the belt. Make two rows of holes in each end of the belt, so that the holes in each row will alternate with those in preceding row, making a zigzag. Four holes will be required for a three-inch belt in each end, two holes in each row; in a six-inch belt, place seven holes in each end, four in the row nearest the end.

To find the length of a belt when the exact length cannot be measured conveniently, measure a straight line from the center of one pulley to the center of the other. Add together half the diameter of each pulley, and multiply that by $3\frac{1}{4}$ (3.1416). The result added to twice the distance between the centers will give the total length of the belt.

A belt will work best if it is allowed to sag just a trifle.

The seam side of a rubber belt should be placed outward, or away from the pulley.

If such a belt slips, coat the inside with boiled linseed oil or soap.

Cotton belting may be preserved by painting the pulley side while running with common paint, afterward applying soft oil or grease.

If a belt slips apply a little oil or soap to the pulley side.

Q. How does the capacity of belts vary?

A. In proportion to width and also to the speed. Double the width and you double the capacity; also, within a certain limit, double the speed and you double the capacity. A belt should not be run over 5,000 feet per minute. One four-inch belt will have the same capacity as two two-inch belts.

Q. How are piston rods and valve rods packed so that the steam cannot escape around them?

A. By packing placed in stuffing-boxes. The stuffing is of some material that has a certain amount of elasticity, such as lamp wick, hemp, soap stone, etc., and certain patent preparations. The packing is held in place by a gland, as it is called, which acts to tighten the packing as the cap of the stuffing-box is screwed up.

Q. How would you repack a stuffing-box?

A. First remove the cap and the gland, and with a proper tool take out all the old packing. Do not use any rough instrument like a file, which is liable to scratch the rod, for any injury to the smooth surface of the rod will make it leak steam or work hard.

If patent packing is used, cut off a sufficient number of lengths to make the required rings. They should be exactly the right length to go around inside the stuffing-box. If too long, they cannot be screwed up tight, as the ends will press together and cause irregularities. If too short, the ends will not meet and will leak steam. Cut the ends diagonally so that they will make a lap joint instead of a square one. When the stuffing-box has been filled, place the gland in position and screw up tight. Afterwards loosen the nuts a trifle, as the steam will cause the packing to expand, usually. The stuffing-box should be just as loose as it can be and not allow leakage of steam. If steam leaks, screw up the box a little tighter. If it still leaks, do not screw up as tight as you possibly can, but repack the box. If the stuffing-box is too tight, either for the piston rod or valve steam, it will cause the engine to work hard, and may groove the rods and spoil them.

If hemp packing is used, pull the fibres out straight and free, getting rid of all knots and lumps. Twist together a few of the fibres, making three cords, and braid these three cords together and soak them with oil or grease, wind around the rod till stuffing-box is sufficiently full, replace the gland, and screw up as before.

Stuffing-box for water piston of pump may be packed as described above, but little oil or grease will be needed.

Never pack the stuffing-box too tight, or you may flute the rod and spoil it.

Always keep the packing in a clean place, well covered up, never allowing any dust to get into it, for the dust or grit is liable to cut the rod.

CHAPTER X.

It is something to be able to run a farm engine and keep out of trouble. It is even a great deal if everything runs smoothly day in and day out, if the engine looks clean, and you can always develop the amount of power you need. You must be able to do this before you can give the fine points of engineering much consideration.

When you come to the point where you are always able to keep out of trouble, you are probably ready to learn how you can make your engine do more work on less fuel than it does at present. In that direction the best of us have an infinite amount to learn. It is a fact that in an ordinary farm engine only about 4 per cent of the coal energy is actually saved and used for work; the rest is lost, partly in the boiler, more largely in the engine. So we see what a splendid chance there is to save.

If we are asked where all the lost energy goes to, we might reply in a general sort of way, a good deal goes up the smokestack in smoke or unused fuel; some is radiated from the boiler in the form of heat and is lost without producing any effect on the steam within the boiler; some is lost in the cooling of the steam as it passes to the steam cylinder; some is lost in the cooling of the cylinder itself after each stroke; some is lost through the pressure on the back of the steam valve, causing a friction that requires a good deal of energy in the engine to overcome; some is lost in friction in the bearings, stuffing-boxes, etc. At each of these points economy may be practiced if the engineer knows how to do it. We offer a few suggestions.

THEORY OF STEAM POWER.

As economy is a scientific question, we cannot study it intelligently without knowing something of the theory of heat, steam and the transmission of power. There will be nothing technical in the following pages; and as soon as the theory is explained in simple language, any intelligent person will know for himself just what he ought to do in any given case.

First, let us define or describe heat according to the scientific theory. Scientists suppose that all matter is made up of small particles called molecules, so small that they have never been seen. Each molecule is made up of still smaller particles called atoms. There is nothing smaller than an atom, and there are only about sixty-five different kinds of atoms, which are called elements; or rather, any substance made up of only one kind of atom is called an element. Thus iron is an element, and so is zinc, hydrogen, oxygen, etc. But a substance like water is not an element, but a compound, since its molecules are made up of an atom of oxygen united with two atoms of hydrogen. Wood is made up of many different kinds of atoms united in various ways. Air is not a compound, but a mixture of oxygen, nitrogen and a few other substances in small quantities.

The reason why air is a mixture and not a compound is an interesting one, and brings us to our next point. In order to form a compound, two different kinds of atoms must have an attraction for each other. There is no attraction between oxygen and nitrogen; but there is great attraction between oxygen and carbon, and when they get a chance they rush together like long separated overs. Anthracite coal is almost pure carbon. So is charcoal. Soft coal consists of carbon with which various other things are united, one of them being hydrogen. This is interesting and important, because it accounts for a curious thing in firing up boilers with soft coal. We have already said that water is oxygen united with hydrogen. When soft coal burns, not only does the carbon unite with oxygen, but the hydrogen unites with oxygen and forms water, or steam. While the boilers are cold they will condense the water or steam in the smoke, just as a cold plate in a steamy room will condense water from the steamy air, so sweating.

Now the scientists suppose that two or three atoms

stick together by reason of their attraction for each other and form molecules. These molecules in turn stick together and form liquids and solids. The tighter they stick, the harder the substance. At the same time, these molecules are more or less loose, and are constantly moving back and forth. In a solid like iron they move very little; but a current of electricity through iron makes the molecules move in a peculiar way. In a liquid like water, the molecules cling together very loosely, and may easily be pulled apart. In any gas, like air or steam, the molecules are entirely disconnected, and are constantly trying to get farther apart.

Heat, says the scientist, is nothing more or less than the movement of the molecules back and forth. Heat up a piece of iron in a hot furnace, and the molecules keep getting further and further apart, and the iron gets softer and softer, till it becomes a liquid. If we take some liquid like water and heat it, the molecules get farther and farther apart, till the water boils, as we say, or turns into steam. As steam the molecules have broken apart entirely, and are beating back and forth so rapidly that they have a tendency to push each other farther and farther apart. This pushing tendency is the cause of steam pressure. It also explains why steam has an expansive power.

Heat, then, is the movement of the molecules back and forth. There are three fixed ranges in which they move; the small range makes a solid; the next range makes a liquid; the third range makes a gas, such as steam. These three states of matter as affected by heat are very sharp and definite. The point at which a solid turns to a liquid is called the melting point. The melting point of ice is 32° Fahr. The point at which it turns to a gas is called the boiling point. With water that is 212° Fahr. The general tendency of heat is to push apart, or expand; and when the heat is taken away the substances contract.

Let us consider our steam boiler. We saw that some different kinds of atoms have a strong tendency to rush together; for example, oxygen and carbon. The air is full of oxygen, and coal and wood are full of carbon. When they are raised to a certain temperature, and the molecules get loose enough so that they can tear themselves away from whatever they are attached to, they rush together with terrible force, which sets all surrounding molecules to vibrating faster than ever. This means that heat is given out.

Another important thing is that when a solid changes to a liquid, or a liquid to a gas, it must take up a certain amount of heat to keep the molecules always just so far apart. That heat is said to become latent, for it will not show in a thermometer, it will not cause anything to expand, nor will it do any work. It merely serves to hold the molecules just so far apart.

HOW ENERGY IS LOST.

We may now see some of the ways in which energy is lost. First, the air which goes into the firebox consists of nitrogen as well as oxygen. That nitrogen is only in the way, and takes heat from the fire, which it carries out at the smokestack.

Again, if the air cannot get through the bed of coals easily enough, or there is not enough of it so that every atom of carbon, etc., will find the right number of atoms of oxygen, some of the atoms of carbon will be torn off and united with oxygen, and the other atoms of carbon, left without any oxygen to unite with, will go floating out at the smokestack as black smoke. Also, the carbon and the oxygen cannot unite except at a certain temperature, and when fresh fuel is thrown on the fire it is cold, and a good many atoms of carbon after being loosened up, get cooled off again before they have a chance to find an atom of oxygen, and so they, too, go floating off and are lost.

If the smoke could be heated up, and there were enough oxygen mixed with it, the loose carbon would still burn and produce heat, and there would be an economy of fuel. This has given rise to smoke consumers, and arranging two boilers, so that when one is being fired the heat from the other will catch the loose carbon before it gets away and burn it up.

So we have these points:

1. Enough oxygen or air must get into a furnace so

that every atom of carbon will have its atom of oxygen. This means that you must have a good draft and that the air must have a chance to get through the coal or other fuel.

2. The fuel must be kept hot enough all the time so that the carbon and oxygen can unite. Throwing on too much cold fuel at one time will lower the heat beyond the economical point and cause loss in thick smoke.

3. If the smoke can pass over a hot bed of coals, or through a hot chamber, the carbon in it may still be burned. This suggests putting fuel at the front of the firebox, a little at a time, so that its smoke will have to pass over a hot bed of coals and the waste carbon will be burned. When the fresh fuel gets heated up, it may be pushed farther back.

From a practical point of view these points mean, No dead plates in a furnace to keep the air from going through coal or wood; a thin fire so the air can get through easily; place the fresh fuel where its smoke will have a chance to be burned; and do not cool off the furnace by putting on much fresh fuel at a time.

(Later we will give more hints on firing.)

HOW HEAT IS DISTRIBUTED.

We have described heat as the movement of molecules back and forth at a high rate of speed. If these heated molecules beat against a solid like iron, its molecules are set in motion, one knocks the next, and so on, just as you push one man in a crowd, he pushes the next, and so on till the push comes out on the other side. So heat passes through iron and appears on the other side. This is called "conduction."

All space is supposed to be filled with a substance in which heat, light, etc., may be transmitted, called the ether. When the molecules of a sheet of iron are heated, or set vibrating, they transmit the vibration through the air, or ether. This is called "radiation." Heat is "conducted" through solid and liquid substances, and "radiated" through gases.

Now some substances conduct heat readily, and some do so with the greatest difficulty. Iron is a good conductor; carbon, or soot on the flues of a boiler, and lime or scale on the inside of a boiler, are very poor conductors. So the heat will go through the iron and steel to the water in a boiler quickly and easily, and a large per cent of the heat of the furnace will get to the water in a boiler. When a boiler is old and is clogged with soot and coated with lime, the heat cannot get through easily, and goes off in the smokestack. The air coming out of the smokestack will be much hotter; and that extra heat is lost.

Iron is a good radiator, too. So if the outer shell of a boiler is exposed to the air, a great deal of heat will run off into space and be lost. Here, then, is where you need a non-conductor, as it is called, such as lime, wood, or the like.

Economy says, cover the outside of a boiler shell with a non-conductor. This may be brickwork in a set boiler; in a traction boiler it means a jacket of wood, plaster, hair, or the like. The steam pipe, if it passes through outer air, should be covered with felt; and the steam cylinder ought to have its jacket, too.

At the same time all soot and all scale should be scrupulously cleaned away.

PROPERTIES OF STEAM.

As we have already seen, steam is a gas. It is slightly blue in color, just as the water in the ocean is blue, or the air in the sky.

We must distinguish between steam and vapor. Vapor is small particles of water hanging in the air. They seem to stick to the molecules composing the air, or hang there in minute drops. Water hanging in the air is, of course, water still. Its molecules do not have the movement that the molecules of a true gas do, such as steam is. Steam, moreover, has absorbed latent heat, and has expansive force; but vapor has no latent heat, and no expansive force. So vapor is dead and lifeless, while steam is live and full of energy to do work.

When vapor gets mixed with steam it is only in the way; it is a sort of dead weight that must be carried;

and the steam power is diminished by having vapor mixed with it.

Now all steam as it bubbles up through water in boiling takes up with it a certain amount of vapor. Such steam is called "wet" steam. When the vapor is no longer in it, the steam is called "dry" steam. It is dry steam that does the best work, and that every engineer wants to get.

While water will be taken up to great heights in the air and form clouds, in steam it will not rise very much, and at a certain height above the level of the water in a boiler the steam will be much drier than near the surface. For this reason steam domes have been devised, so that the steam may be taken out at a point as high as possible above the water in the boiler, and so be as dry as possible. Also "dry tubes" have been devised, which let the steam pass through many small holes that serve to keep back the water to a certain extent.

However, there will be more or less moisture in all steam until it has been superheated, as it is called. This may be done by passing it through the hot part of the furnace, where the added heat will turn all the moisture in the steam into steam, and we shall have perfectly dry steam.

The moment, however, that steam goes through a cold pipe, or one cooled by radiation, or goes into a cold cylinder, or a cylinder cooled by radiation, some of the steam will turn to water, or condense, as it is called. So we have the same trouble again.

Much moisture passing into the cylinder with the steam is called "priming." In that case the dead weight of water has become so great as to kill a great part of the steam power.

HOW TO USE THE EXPANSIVE POWER OF STEAM.

We have said that the molecules in steam are always trying to get farther and farther apart. If they are free in the air, they will soon scatter; but if they are confined in a boiler or cylinder they merely push out in every direction, forming "pressure."

When steam is let into the cylinder it has the whole accumulated pressure in the boiler behind it, and of course that exerts a strong push on the piston. Shut off the boiler pressure and the steam in the cylinder will still have its own natural tendency to expand. As the space in the cylinder grows larger with the movement of the piston from end to end, the expansive power of the steam becomes less and less, of course. However, every little helps, and the push this lessened expansive force exerts on the piston is so much energy saved. If the full boiler pressure is kept on the piston the whole length of the stroke, and then the exhaust port is immediately opened, all this expansive energy of the steam is lost. It escapes through the exhaust nozzle into the smokestack and is gone. Possibly it cannot get out quickly enough, and causes back pressure on the cylinder when the piston begins its return stroke, so reducing the power of the engine.

To save this the skilled engineer "notches up" his reverse lever, as they say. The reverse lever controls the valve travel. When the lever is in the last notch the valve has its full travel. When the lever is in the center notch the valve has no travel at all, and no steam can get into the cylinder; on the other side the lever allows the valve to travel gradually more and more in the opposite direction, so reversing the engine.

As the change from one direction to the other direction is, of course, gradual, the valve movement is shortened by degrees, and lets steam into the cylinder for a correspondingly less time. At its full travel it perhaps lets steam into the cylinder for three-quarters of its stroke. For the last quarter the work is done by the expansive power of the steam.

Set the lever in the half notch, and the travel of the valve is so altered that steam can get into the cylinder only during half the stroke of the piston, the work during the rest of the stroke being done by the expansive force of the steam.

Set the lever in the notch next to the middle notch, or the quarter notch, and steam will get into the cylinder only during a quarter of the stroke of the piston, the work being done during three-quarters of the stroke by the expansive force of the steam.

Obviously the more the steam is expanded the less work it can do. But when it escapes at the exhaust there will be very little pressure to be carried away and lost.

Therefore when the load on his engine is light the economical engineer will "notch up" his engine with the reverse lever, and will use up correspondingly less steam and save correspondingly more fuel. When the load is unusually heavy, however, he will have to use the full power of the pressure in the boiler, and the waste cannot be helped.

THE COMPOUND ENGINE.

The compound engine is an arrangement of steam cylinders to save the expansive power of steam at all times by letting the steam from one cylinder where it is at high pressure into another after it exhausts from the first, in this second cylinder doing more work purely by the expansive power of the steam.

The illustration shows a sectional view of a compound engine having two cylinders, one high pressure and one low. The low pressure cylinder is much larger than the high pressure. There is a single plate between them called the center head, and the same piston rod is fitted with two pistons, one for each cylinder. The steam chest does not receive steam from the boiler, but from the exhaust of the high pressure cylinder. The steam from the boiler goes into a chamber in the double valve, from which it passes to the ports of the high pressure cylinder. At the return stroke the exhaust steam escapes into the steam chest, and from there it passes into the low pressure cylinder. There may be one valve riding on the back of another; but the simplest form of compound engine is built with a single double valve, which opens and closes the ports for both cylinders at one movement.

Theoretically the compound engine should effect a

genuine economy. In practice there are many things to operate against this. Of course if the steam pressure is low to start with, the amount of pressure lost in the exhaust will be small. But if it is very high, the saving

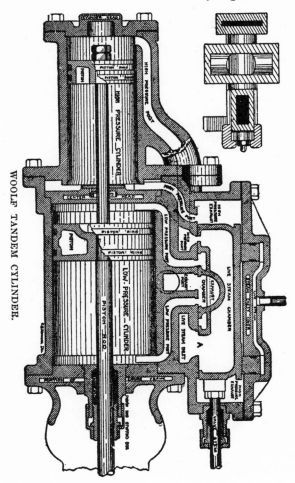

in the low pressure cylinder will be relatively large. If the work can be done just as well with a low pressure, it would be a practical waste to keep the pressure abnormally high in order to make the most of the compound engine.

An engine must be a certain size before the saving of a compound cylinder will be appreciable. In these days nearly all very large engines are compound, while small engines are simple.

Another consideration to be taken into account is that a compound is more complicated and so harder to manage ; and when any unfavorable condition causes loss it causes proportionately more loss on a compound than on a simple engine. For these and other reasons compound engines have been used less for traction purposes than simple engines have. It is probable that a skilled and thoroughly competent engineer, who would manage his engine in a scientific manner, would get more out of a compound than out of a simple ; and this would be especially true in regions where fuel is high. If fuel is cheap and the engineer unskilled, a compound engine would be a poor economizer.

FRICTION.

We have seen that the molecules of water have a tendency to stick in the steam as vapor or moisture. All molecules that are brought into close contact have more or less tendency to stick together, and this is called friction. The steam as it passes along the steam pipe is checked to a certain extent by the friction on the sides of the pipe. Friction causes heat, and it means that the heat caused has been taken from some source of energy. The friction of the steam diminishes the energy of the steam.

So, too, the fly wheel moving against the air suffers friction with the air, besides having to drive particles of air out of its path. All the moving parts of an engine where one metal moves on another suffer friction, since where the metals are pressed very tightly together they have more tendency to stick than when not pressed so tightly When iron is pressed too tightly, as under the blows of a hammer in a soft state, it actually welds together solidly.

There is a great deal of friction in the steam cylinder, since the packing rings must press hard against the walls of the cylinder to prevent the steam from getting through. There is a great deal of friction between the D valve and its seat, because of the high steam pressure on the back of the valve. There is friction in the stuffing boxes both of the valve and the piston. There is friction at all the bearings.

There are various ways in which friction may be reduced. The most obvious is to adjust all parts so nicely that they will bind as little as possible. The stuffing-boxes will be no tighter than is necessary to prevent leaking of steam; and so with the piston rings. Journal boxes will be tight enough to prevent pounding, but no tighter. To obtain just the right adjustment requires great patience and the keen powers of observation and judgment.

The makers of engines try to reduce friction as much as possible by using anti-friction metals in the boxes. Iron and steel have to be used in shafts, gears, etc., because of the strength that they possess; but there are some metals that stick to each other and to iron and steel much less than iron or steel stick to each other when pressed close together. These metals are more or less soft; but they may be used in boxes and journal bearings. They are called anti-friction metals. The hardest for practical purposes is brass, and brass is used where there is much wear. Where there is less wear various alloys of copper, tin, zinc, etc., may be used in the boxes. One of these is babbit metal, which is often used in the main journal box.

All these anti-friction metals wear out rapidly, and they must be put in so that they can be adjusted or renewed easily.

But the great anti-friction agent is oil.

Oil is peculiar in that while the molecules seem to stick tightly together and to a metal like iron or steel, they roll around upon each other with the utmost ease. An ideal lubricator is one that sticks so tight to the journal that it forms a sort of cushion all around it, and prevents any of its molecules coming into contact with the molecules of the metal box. All the friction then takes place between the different molecules of oil, and this friction is a minimum.

The same principle has been applied to mechanics in the ball bearing. A number of little balls roll around between the journal and its box, preventing the two metals from coming into contact with each other; while the balls, being spheres, touch each other only at a single point, and the total space at which sticking can occur is reduced to a minimum.

As is well known, there is great difference in oils. Some evaporate, like gasoline and kerosene, and so disappear quickly. Others do not stick tightly to the journal, so are easily forced out of place, and the metals are allowed to come together. What is wanted, then, is a heavy, sticky oil that will not get hard, but will always form a good cushion between bearings.

Steam cylinders cannot be oiled directly, but the oil must be carried to the steam chest and cylinder in the steam. A good cylinder oil must be able to stand a high temperature. While it is diffused easily in the steam, it must stick tightly to the walls of the steam cylinder and to the valve seat, and keep them lubricated. Once it is stuck to the metal, the heat of the steam should not evaporate it and carry it away.

Again, a cylinder oil should not have any acid in it which would have a tendency to corrode the metal. Nearly all animal fats do have some such acid. So tallow and the like should not be placed where they can corrode iron or steel. Lard and suet alone are suitable for use on an engine.

When it comes to lubricating traction gears, other problems appear. A heavy grease will stick to the gears and prevent them from cutting; but it will stick equally to all sand and grit that may come along, and that, working between the cogs, may cut them badly. So some engineers recommend the use on gears of an oil that does not gather so much dirt.

The friction of the valve on its seat due to the pressure of the steam on its back has given rise to many inventions for counteracting it. The most obvious of these is what is called "the balanced valve." In the compound engine, where the steam pressure is obtained upon both sides of the valve, it rides much more lightly on its seat—so lightly, indeed, that when steam pressure is low, as in going down hill or operating under a light load, plunger pistons must be used to keep the valve down tight on its seat.

The poppet valves were devised to obviate the undue friction of the D valve; but the same loss of energy is to a certain extent transferred, and the practical saving is not always equal to the theoretical. On large stationary engines rotary valves and other forms, such as are used on the Corliss engine, have come into common use; but they are too complicated for a farm engine, which must be as simple as possible, with least possible liability of getting out of order.

CHAPTER XI.

PRACTICAL POINTS.

The first practical point in the direction of farm engine economy is to note that the best work can be done only when every part of the engine and boiler are in due proportion. If the power is in excess of the work to be done, there is loss; if the grate surface is too large cold air gets through the fuel and prevents complete combustion, and if the grate surface is too small, not enough air gets in; if the steaming power of the boiler is too large, heat is radiated away that otherwise could be saved, for every foot of exposed area in the boiler is a source of loss; if the steaming power of the boiler is too low for the work to be done, it requires extra fuel to force the boiler to do its work, and any forcing means comparatively large loss or waste. It will be seen that not only must the engine and boiler be built with the proper proportions, but they must be bought with a nice sense of proportion to the work expected of them. This requires excellent judgment and some experience in measuring work in horsepowers.

GRATE SURFACE AND FUEL.

The grate surface in a firebox should be not less than two-thirds of a square foot per horsepower, for average size traction engines. If the horsepower of an engine is small, proportionately more grate surface will be needed; if it is large, the grate surface may be proportionately much smaller. An engine boiler 7x8x200 rev., with 100 lbs. pressure, should have a great surface not less than six square feet, and seven would be better. In a traction engine there is always a tendency to make the grate surface as small as possible, so that the engine will not be cumbersome.

Another reason why the grate surface should be sufficiently large is that forced draft is a bad thing, since it has a tendency to carry the products of combustion and hot gases through the smokestack and out into space before they have time to complete combustion and especially before the heat of the gases has time to be absorbed by the boiler surface. A large grate surface, then, with a moderate draft, is the most economical.

The draft depends on other things, however. If a great deal of fine fuel is thrown on a fire, the air must be forced through, because it cannot get through in the natural way. This results in waste. So a fire should be as open as possible. Coal should be "thin" on the grates; wood should be thrown in so that there will be plenty of air spaces; straw should be fed in just so that it will burn up completely as it goes in. Moderate size coal is better than small or fine. Dust in coal checks the draft. A good engineer will choose his fuel and handle his fire so that he can get along with as little forced draft as possible.

In a straw burning engine a good circulation of air can be obtained, if the draft door is just below the straw funnel, by extending the funnel into the furnace six inches or so. This keeps the straw from clogging up the place where the air enters and enables it to get at the fuel so much more freely that the combustion is much more complete.

We have already suggested that in firing with coal, the fresh fuel be deposited in front, so that the smoke will have to pass over live coals and so the combustion will be more complete. Then when the coal is well lighted it can be poked back over the other portions of the grate. This method has another advantage, in that the first heating is usually sufficient to separate the pure coal from the mineral substances which form clinkers, and most of the clinkers will be deposited at that one point in the grate. Here they can easily be lifted out, and will not seriously interfere with the burning of the coal as they would if scattered all over the grate. Clinkers in front can easily be taken out by hooking the poker over them toward the back of the firebox and pulling them up and to the front.

They often come out as one big mass which can be easily lifted out.

The best time to clean the grate is when there is a good brisk fire. Then it will not cause steam to go down. Stirring a fire does little good. For one thing, it breaks up the clinkers and allows them to run down on the grate bars when they stick and finally warp the bars. If the fire is not stirred the clinkers can be lifted out in large masses. Stirring a fire also creates a tendency to choke up or coke, and interferes with the even and regular combustion of the coal at all points.

The highest heat that can be produced is a yellow heat. When there is a good yellow heat, forced draft will only carry off the heat and cause waste. It will not cause still more rapid combustion. When the heat is merely red, increased draft will raise the temperature. Combustion is not complete until the flame shows yellow. However, if the draft is slight and time is given, red heat will be nearly as effective, but it will not carry the heated gases over so large a part of the heating surface of the boiler. With a very large grate surface, red heat will do very well. Certainly it will be better than a forced draft, or an effort at heating beyond the yellow point.

BOILER HEATING SURFACE.

The heat of the furnace does its work only as the heated gases touch the boiler surface. The iron conducts the heat through to the water, which is raised to the boiling point and turned into steam.

Now the amount of heat that the boiler will take up is directly in proportion to the amount of exposed surface and to the time of exposure. If the boiler heating surface is small, and the draft is forced so that the gases pass through rapidly, they do not have a chance to communicate much heat.

Also if the heating surface is too large, so that it cannot all be utilized, the part not used becomes a radiating surface, and the efficiency of the boiler is impaired.

Practice has shown that the amount of heating surface practically required by a boiler is 12 to 15 square feet per horsepower. In reckoning heating surface, all area which the heated gases touch is calculated.

Another point in regard to heating surface in the production of steam is this, that only such surface as is exposed to a heat equal to turning the water into steam is effective. If there is a pressure of 150 lbs. the temperature at which the water would turn to steam would be 357 degrees, and any gases whose temperature was below 357 degrees would have no effect on the heating surface except to prevent radiation. Thus in a return flue boiler the heated gases become cooled often to such an extent before they pass out at the smokestack that they do not help the generation of steam. Yet a heat just below 357 degrees would turn water into steam under 149 lbs. pressure. Though it has work in it, the heat is lost.

Another practical point as to economy in large heating surface is that it costs money to make, and is cumbersome to move about. It may cost more to move a traction engine with large boiler from place to place than the saving in fuel would amount to. So the kind of roads and the cost of fuel must be taken into account and nicely balanced.

However, it may be said that a boiler with certain outside dimensions that will generate 20 horsepower will be more economical than one of the same size that will generate only 10 horsepower. In selecting an engine, the higher the horsepower for the given dimensions, the more economical of both fuel and water.

The value of heating surface also depends on the material through which the heat must penetrate, and the rapidity with which the heat will pass. We have already pointed out that soot and lime scale permit heat to pass but slowly and if they are allowed to accumulate will greatly reduce the steaming power of a boiler for a given consumption of fuel. Another point is that the thinner the iron or steel, the better will the heat get through even that. So it follows that flues, being thinner, are better conductors than the sides of the firebox. Long flues are better than short ones in that the long ones allow less soot, etc., to accumulate than the short ones do, and afford more time for the boiler to absorb the heat of the gases.

Again, we have stated that heating surface is valuable only as it is exposed to the gases at a sufficiently high temperature. Some boilers have a tendency to draw the hot gases most rapidly through the upper flues, while the lower flues do not get their proportion of the heat. This results in a loss, for the heat to give its full benefit should be equally distributed.

To prevent the heat being drawn too rapidly through upper flues, a baffle plate may be placed in the smoke box just above the upper flues, thus preventing them from getting so much of the draft.

Again, if the exhaust nozzle is too low down, the draft through the lower flues may be greater than through the upper. This is remedied by putting a piece of pipe on the exhaust to raise it higher in the smokestack.

EXPANSION AND CONDENSATION.

We have already pointed out that economy results if we hook up the reverse lever so that the expansive force of the steam has an opportunity to work during half or three-quarters of the stroke.

One difficulty arising from this method is that the walls of the cylinder cool more rapidly when not under the full boiler pressure. Condensation in the cylinder is a practical difficulty which should be met and overcome as far as possible.

High speed gives some advantage. A judicious use of cushion helps condensation somewhat also, because when any gas like steam or air is compressed, it gives off heat, and this heat in the cushion will keep up the temperature of the cylinder. This cannot be carried very far, however, for the back pressure of cushion will reduce the energy of the engine movement.

LEAD AND CLEARANCE.

Too much clearance will detract from the power of an engine, as there is just so much more waste space to be filled with hot steam. Too little clearance will cause pounding.

Likewise there will be loss of power in an engine if the lead is too great or too little. The proper amount of lead differs with conditions. A high speed engine requires more than a low speed, and if an engine is adjusted for a certain speed, it should be kept uniformly at that speed, as variation causes loss. The more clearance an engine has the more lead it needs. Also the quicker the valve motion, the less lead required. Sometimes when a large engine is pulling only a light load and there is no chance to shorten the cut-off, a turn of the eccentric disk for a trifle more lead will effect some economy.

Cut-off should be as sharp as possible. A slow cut-off in reducing pressure before cut-off is complete, causes a loss of power in the engine.

THE EXHAUST.

If the exhaust from the cylinder does not begin before the piston begins its return stroke, there will be back pressure due to the slowness with which the valve opens. The exhaust should be earlier in proportion to the slowness of the valve motion, and also in proportion to the speed of the engine, since the higher the speed the less time there is for the steam to get out. It follows that an engine whose exhaust is arranged for a low speed cannot be run at a high speed without causing loss from back pressure.

In using steam expansively the relative proportion between the back pressure and the force of the steam is of course greater. So in using steam expansively the back pressure must be at a minimum, and this is especially true in the compound engine. So many things affect this, that it becomes one of the reasons why it is hard to use a compound engine with as great economy as theory would indicate.

Another thing, the smallness of the exhaust nozzle in the smokestack affects the back pressure. The smaller the nozzle, the greater the draft a given amount of steam will create; but the more back pressure there will be, due to the inability of the exhaust steam to get out easily. So the exhaust nozzle should be as large as circumstances will permit. It is a favorite trick with engineers testing the pulling power of their engines to remove the exhaust nozzle entirely for a few minutes when the fire is up. The back pressure saved will at once show in the pulling power of the engine, and every one will be surprised. Of course the fire couldn't be kept going long without the nozzle on. We have already pointed out that a natural draft is better than a forced one. Here is another reason for it.

LEAKS.

Leaks always cause a waste of power. They may usually be seen when about the boiler; but leaks in the piston and valve will often go unnoticed.

It is to be observed that if a valve does not travel a short distance beyond the end of its seat, it will wear the part it does travel on, while the remaining part will not wear and will become a shoulder. Such a shoulder will nearly always cause a leak in the valve, and besides will add the friction, and otherwise destroy the economy of the engine.

Likewise the piston will wear part of the cylinder and leave a shoulder at either end if it does not pass entirely beyond the steam-tight portion of the inside of the cylinder. That it may always do this and yet leave sufficient clearance, the counterbore has been devised. All good engines are bored larger at each end so that the piston will pass beyond the steam-tight portion a trifle at the end of each stroke. Of course it must not pass far enough to allow any steam to get through.

Self-setting piston rings are now generally used. They are kept in place by their own tension. There will always be a little leakage at the lap. The best lap is probably a broken joint rather than a diagonal one. Moreover, as the rings wear they will have a tendency to get loose unless they are thickest at a point just opposite to the lap, since this is the point at which it is necessary to make up for the tension lost by the lapping.

CHAPTER XII.

DIFFERENT TYPES OF ENGINES.

STATIONARY.

So far we have described and referred exclusively to the usual form of the farm traction engine, which is nearly always the simplest kind of an engine, except in one particular, namely, the reverse which gives a variable cut-off. Stationary engines, however, are worked under such conditions that various changes in the arrangement

D. JUNE & CO.'S STATIONARY FOUR-VALVE ENGINE.

may be made which gives economy in operating, or other desirable qualities. We will now briefly describe some of the different kinds of stationary engines.

THROTTLING AND AUTOMATIC CUT-OFF TYPES.

Engines may be divided into two classes, namely, throttling and automatic cut-off engines. The throttling engine regulates the speed of the engine by cutting off the supply of steam from the boiler, either by the hand of the engineer on the throttle or by a governor working a special throttling governor valve. Railroad locomotives are throttling engines, and moreover they have no governor, the speed being regulated by the engineer at the throttle valve. Traction engines are usually throttling engines provided with a governor.

An automatic cut-off engine regulates its speed by a governor connected with the valve, and does it by shortening the time during which steam can enter the cylinder. This is a great advantage, in that the expansive power of steam is given a chance to work, while in the throttling engine steam is merely cut off. The subject has been fully discussed under "Economy in Running a Farm Engine." An automatic cut-off engine is much the most economical.

While on traction engines the governor is usually of the ball variety, on stationary engines improved forms of governors are also placed in the fly wheel, and work in various ways, according to the requirements of the valve gear.

THE CORLISS ENGINE.

The Corliss engine is a type now well known and made by many different manufacturers. It is considered one of the most economical stationary engines made, but cannot be used for traction purposes. It may be compound, and may be used with a condenser. It cannot be used as a high speed engine, since the valves will not work rapidly enough.

The peculiarity of a Corliss engine is the arrangement of the valves. It has four valves instead of one, and they are of the semi-rotary type. They consist of a small, long cylinder which rocks back and forth, so as to close and open the port, which is rather wide and short compared to other types. There is a valve at each end of the cylinder opening usually into the clearance space, to admit steam; and two more valves below the cylinder for the exhaust. These exhaust valves allow any water of condensation to run out of the cylinder. Moreover, as the steam when it leaves the cylinder is much colder than when it enters, the exhaust always cools the steam ports, and when the same ports are used for exhaust and admission the fresh steam has to pass through ports that have been cooled and cause condensation. In the Corliss engine the exhaust does not have an opportunity to cool the live steam ports and the condensation is reduced. This works considerable economy.

Also the Corliss valves have little friction from steam pressure on their own backs, since the moment they are lifted from their seats they work freely. The valves are controlled by a governor so as to make the automatic cut-off engine.

The Corliss type of frame for engine is often used on traction engines and means the use of convex shoes on cross-head and concave ways or guides. In locomotive type, cross-head slides in four square angle guides.

THE HIGH SPEED ENGINE.

A high speed engine means one in which the speed of the piston back and forth is high, rather than the speed of rotation, there being sometimes a difference. High speed engines came into use because of the need of such to run dynamos for electric lighting. Without a high speed engine an intermediate gear would have to be used, so as to increase the speed of the operating shaft. In the high speed engine this is done away with.

As an engine's power varies directly as its speed as well as its cylinder capacity or size, an engine commonly used for ten horsepower would become a twenty horsepower engine if the speed could be doubled. So high speed engines are very small and compact, and require less metal to build them. Therefore they should be much cheaper per horsepower.

A high speed engine differs from a low speed in no essential particular, except the adjustment of parts. A high steam pressure must be used; a long, narrow valve port is used, so that the full steam pressure may be let on quickly at the beginning of the stroke when the piston is reversing its motion and needs power to get started quickly on its return; the slide valve must be

used, since the semi-rotary Corliss would be too wide and
short for a quick opening. Some high speed engines are
built which use four valves, as does the Corliss. The
friction of the slide valve is usually "balanced" in some
way, either by "pressure plates" above the valve, which
prevent the steam from getting at the top and pressing
the valve down, or by letting the steam under the valve,
making it slide on narrow strips, since the pressure above
would then be reduced in proportion with the smallness
of the bearing surface below, and if the bearing surface
were very small the pressure above would be correspondingly small, perhaps only enough to keep the valve in
place. Some automatic cut-off gear is almost always
used. A high speed engine may attain 900 revolutions
per minute, 600 being common. In many ways it is
economical.

CONDENSING AND NON-CONDENSING.

In the traction engine the exhaust is used in the smokestack to help the draft, since the smokestack must neces-
sarily be short. A stationary engine is usually provided
with a boiler set in brickwork, and a furnace with a high
chimney, which creates all the draft needed. In other
words, the heated gases wasted in a traction engine are
utilized to make the draft.

It then becomes desirable to save the power in the exhaust steam in some way. Some of this can be used to
heat the feed water, but only a fraction of it.

Now when the exhaust steam issues into the air it
must overcome the pressure of the atmosphere, nearly
15 lbs. to the square inch, which is a large item to begin
with. This can be saved by letting the steam exhaust
into a condenser, where a spray of cold water or the
like suddenly condenses the steam so that a vacuum is
created. There is then no back pressure on the exhaust
steam, theoretically. Practically a perfect vacuum cannot be created, and there is a back pressure of 2 or 3
lbs. per square inch. By the use of a condenser a back
pressure of about 12 lbs. is taken off the head of the
piston on its return stroke, a matter of considerable

economy. But an immense amount of water is required to run a condenser, namely, 20 times as much for a given saving of power as is required in a boiler to make that

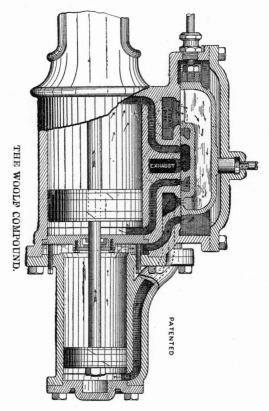

THE WOOLF COMPOUND.

STEAM PIPE

EXHAUST

PATENTED

power. So condensers are used only where water is cheap.

COMPOUND AND CROSS-COMPOUND.

We have already explained the economy effected by the compound engine, in which a large low pressure cylinder is operated by the exhaust from a small high pressure cylinder. In the cut used for illustration the low pressure cylinder is in direct line with the high pressure cylinder, and one piston rod connects both pistons. This arrangement is called the "tandem." Sometimes the low pressure cylinder is placed by the side of the high pressure, or at a distance from it, and operates another piston and connecting rod. By using a steam chest to store the exhaust steam and varying the cut-off of the two cylinders, the crank of the low pressure may be at an angle of 90 degrees with the crank of the high pressure, and there can be no dead center.

When a very high pressure of steam is used the exhaust from the low pressure cylinder may be used to operate a third cylinder; and the exhaust from that to operate a fourth. Engines so arranged are termed triple and quadruple expansion engines, or multiple expansion.

The practical saving of a compound engine when its value can be utilized to the full is 10 per cent to 20 per cent. Small engines are seldom compounded, large engines nearly always.

CHAPTER XIII.

GAS AND GASOLINE ENGINES.

The gas and gasoline engines (they are exactly the same except that one generates the gas it needs from gasoline, while the other takes common illuminating gas, the use of gas or gasoline being interchangeable on the same engine by readjustment of some of the parts) are operated on a principle entirely different from steam. While they are arranged very much as a steam engine, the power is given by an explosion of gas mixed with air in the cylinder. Instead of being a steady pressure like that furnished by steam, it is a sudden pressure given to one end of the piston usually once in four strokes or two revolutions, one stroke being required to draw the gasoline in, the second to compress it, the third to receive the effect of the explosion (this is the only power stroke), the fourth to push out the burned gases preparatory to admitting a new charge. The fact that force is given the cylinder at such wide intervals makes it necessary to have an extra heavy flywheel to keep the engine steady, and the double cylinder engine which can give a stroke at least every revolution is still better and is indispensable when the flywheel cannot be above a certain weight.

For small horsepowers, such as are required for pumping, feed grinding, churning, etc., the gas engine is so much more convenient and so very much cheaper in operation than the small steam engine that it is safe to say that within a very few years the gas engine will have completely displaced the small steam engine. In fact, the discovery of the gas engine permits the same economies for the small engine that the progress in steam engineering has made possible for the large steam engine. As yet the gas engine has made little or no progress against the large steam plant, with its Corliss engine, its triple expansion, its condenser, and all the other appliances which are not practicable with the small engine.

COMPARISON OF STEAM AND GAS ENGINES.

The following points prepared by an experienced farm engine manufacturer will show clearly the advantages of the gas engine over the steam engine for general use about a farm:

In the first place, the farmer uses power, as a rule, at short intervals, and also uses small power. Should he install a steam engine and wish power for an hour or two, it would be necessary for him to start a fire under the boiler and get up steam before he could start the engine. This would take at least an hour. At the end of the run he would have a good fire and good steam pressure, but no use for it, and would have to let the fire die out and the pressure run down. This involves a great waste of water, time and fuel. With a gasoline engine he is always ready and can start to work within a few minutes after he makes up his mind to do so, and he does not have to anticipate his wants in the power line for half a day. Aside from this, in some states, notably Ohio, the law compels any person operating an engine above ten horsepower to carry a steam engineer's license. This does not apply to a gasoline engine.

Again, the gasoline engine is as portable as a traction engine, and can be applied to all the uses of a traction engine and to general farm use all the rest of the year. At little expense it can be fitted up to hoist hay, to pump water, to husk and shell corn, to saw wood, and even by recent inventions to plowing. It is as good about a farm as an extra man and a team of horses.

A gasoline engine can be run on a pint of gasoline per hour for each horsepower, and as soon as the work is done there is no more consumption of fuel and the engine can be left without fear, except for draining off the water in the water jacket in cold weather. A steam engine for farm use would require at least four pounds of coal per horsepower per hour, and in the majority of cases it would be twice that, taking into consideration the amount of fuel necessary to start the fire and that left unburned after the farmer is through with his power. If you know the cost of crude gasoline at your point and the cost of coal, you can easily figure the exact economy of a gasoline engine for your use. To the economy of fuel question may be added the labor or cost of pumping or hauling water.

The only point wherein a farmer might find it advantageous to have a steam plant would be where he is running a dairy and wished steam and hot water for cleansing his creamery machinery. This can be largely overcome by using the water from the jackets which can be kept at a temperature of about 175 degrees, and if a higher temperature is needed he can heat it with the exhaust from the engine. The time will certainly come soon when no farmer will consider himself up to date until he has a gasoline engine.

Some persons unaccustomed to gasoline may wonder if a gasoline engine is as safe as a steam engine. The fact is, they are very much safer, and do not require a skilled engineer to run them. The gasoline tank is usually placed outside the building, where the danger from an explosion is reduced to a minimum. The only danger that may be encountered is in starting the engine, filling the supply tank when a burner near at hand is in flame, etc. Once a gasoline engine is started and is supplied with gasoline, it may be left entirely alone without care for hours at a time without danger and without adjustment.

With a steam engine there is always danger, unless a highly skilled man is watching the engine every moment. If the water gets a little low he is liable to have an explosion; if it gets a little too high he may knock out a cylinder head in his engine; the fire must be fed every few minutes; the grates cleaned. There is always something to be done about a steam engine.

So here is another point of great saving in a gasoline engine, namely, the saving of one man's time. The man who runs the gasoline engine may give nearly all his time to other work, such as feeding a corn-sheller, a fodder chopper, or the like.

Kerosene may also be used in the same way with a special type of gas engine.

The amounts of fuel required of the different kinds possible in a gas engine are compared as follows by Roper:

Illuminating gas, 17 to 20 cubic feet per horsepower per hour.

Pittsburg natural gas, as low as 11 cubic feet.

74° gasoline, known as stove gasoline, one-tenth of a gallon.

Refined petroleum, one-tenth of a gallon.

If a gas producing plant using coal supplies the gas, one pound of coal per horsepower per hour is sufficient on a large engine.

DESCRIPTION OF THE GAS OR GASOLINE ENGINE.

The gas engine consists of a cylinder and piston, piston rod, cross-head, connecting rod, crank and flywheel, very similar to those used in the steam engine.

There is a gas valve, an exhaust valve, and in connection with the gas valve a self-acting air valve. The gas valve and the exhaust valve are operated by lever arm or cam worked from the main shaft, arranged by spiral gear or the like so that it gets one movement for each two revolutions of the main shaft. Such an engine is called "four cycle" (meaning one power stroke to each four strokes of the piston), and works as follows:

As the piston moves forward the air and fuel valves are simultaneously opened and closed, starting to open just as the piston starts forward and closing just as the piston completes its forward stroke. Gas and air are simultaneously sucked into the cylinder by this movement. As the cylinder returns it compresses the charge taken in during the forward stroke until it again reaches back center. The mixture in the Otto engine is compressed to about 70 pounds per square inch. Ignition then takes place, causing the mixture to explode and giving the force from which the power is derived. As the crank again reaches its forward center the piston uncovers a port which allows the greater part of the burnt gases to escape. As the piston comes back, the exhaust valve is opened, enabling the piston to sweep out the remainder of the burnt gases. By the time the crank is on the back center the exhaust valve is closed and the engine is ready to take another charge, having com-

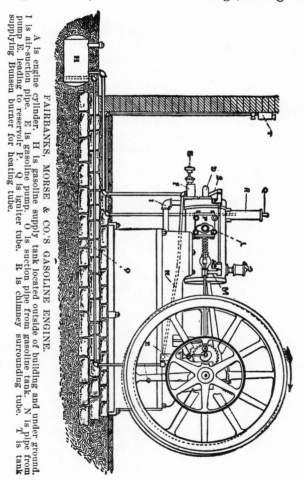

FAIRBANKS, MORSE & CO.'S GASOLINE ENGINE.
A is engine cylinder. H is gasoline supply tank located outside of building and under ground. I is air-suction pipe. E is gasoline pump. O is suction pipe from gasoline tank. N is pipe from pump E. leading to reservoir P. Q is igniter tube. R is chimney surrounding tube. T is tank supplying Bunsen burner for heating tube.

pleted two revolutions or four strokes. The side shaft which performs the functions of opening and closing the valves, getting its motion in the Columbus engine by a pair of spiral gears, makes but one revolution to two of the crank shaft.

Gas engines are governed in various ways. One method is to attach a ball governor similar to the Waters on the steam engine. When the speed is too high, the balls go out, and a valve is closed or partly closed, cutting off the fuel supply. Since the engine takes in fuel only once in four strokes, the governing cannot be so close as on the steam engine, since longer time must elapse before the governor can act.

Another type of governor operates by opening the exhaust port and holding it open. The piston then merely draws in air through the exhaust port, but no gas. This is called the "hit or miss" governing type. One power stroke is missed completely.

The heat caused by the explosion within the cylinder is very great, some say as high as 3,000 degrees. Such a heat would soon destroy the oil used to lubricate the cylinder and make the piston cut, as well as destroying the piston packing. To keep this heat down the cylinder is provided with a water jacket, and a current of water is kept circulating around it to cool it off.

When gas is used, the gas is passed through a rubber bag, which helps to make the supply even. It is admitted to the engine by a valve similar to the throttle valve on an engine.

Gasoline is turned on by a similar valve or throttle. It does not have to be gasefied, but is sucked into the cylinder in the form of a spray. As soon as the engine is started, the high heat of the cylinder caused by the constant explosions readily turns the gasoline to gas as it enters. The supply tank of gasoline is placed outside the building, or at a distance, and stands at a point below the feed. A small pump pumps it up to a small box or feed tank, which has an overflow pipe to conduct any superfluous gasoline back to the supply tank. In the gasoline box or feed tank a conical-shaped basin is filled with gasoline to a certain height, which can be regulated. Whatever this conical basin contains is sucked into the cylinder with the air. By regulating the amount in the basin the supply of gasoline in the cylinder can be regulated to the amount required for any given amount of work. In the Columbus engine this regulation is accomplished by screwing the overflow regulator up or down.

There are two methods of igniting the charge in the cylinder in order to explode it. One is by what is called a gasoline or gas torch. A hollow pin or pipe is fixed in the top of the cylinder. The upper part of this pin or pipe runs up into a gasoline or gas lamp of the Bunsen type where it is heated red hot. When the gas and air in the cylinder are compressed by the back stroke of the piston, some of the mixture is forced up into this pipe or tube until it comes in contact with the heated portion and is exploded, together with the rest of the charge in the cylinder. Of course this tube becomes filled with burnt gases which must be compressed before the explosive mixture can reach the heated portion, and no explosion is theoretically possible until the piston causes compression to the full capacity of the cylinder. The length of the tube must therefore be nicely regulated to the requirements of the particular engine used.

The other method is by an electric spark from a battery. Two electrodes of platinum or some similar substance are placed in the compression end of the cylinder. The spark might be caused by bringing the electrodes sufficiently near together at just the right moment, but the more practical and usual way is to break the current, closing it sharply by means of a lever worked by the gearing at just the moment the piston is ready to return after compressing the charge. The electric spark is by long odds the most desirable method of ignition, being safer and easier to take care of, but it requires some knowledge of electricity and electric connection to keep it always in working order.

OPERATION OF GAS AND GASOLINE ENGINES.

To all intents and purposes the operation of a gas or gasoline engine is the same as that of a steam engine with the care of the boiler eliminated. The care of the engine itself is practically the same, though the bearings are relatively larger in a gasoline or gas engine and do not require adjustment so often. Some manufacturers will tell you that a gas engine requires no attention at all. Any one who went on that theory would soon ruin his engine. To keep a gasoline engine in working order so as to get the best service from it and make it last as long as possible, you should give it the best of care.

An engine of this kind needs just as much oiling and cleaning as a steam engine. All bearings must be lubricated and kept free from dirt, great care must be taken that the piston and cylinder are well lubricated. In addition, the engineer must see that the valves all work perfectly tight, and when they leak in any way they must be taken out and cleaned. Usually the valve seats are cast separate from the cylinder, so that they can be removed and ground when they have worn.

Also the water jacket must be kept in order so that the cylinder cannot become too hot.

STARTING A GASOLINE ENGINE.

It is something of a trick to get a gasoline or gas engine started—especially a gasoline engine—and some skill must be developed in this or there will be trouble. This arises from the fact that when an engine has not been running the cylinder is cold and does not readily gasefy the gasoline. At best only a part of a charge of gasoline can be gasefied, and if the cylinder is very cold indeed the charge will not explode at all till the cylinder is warmed up.

When preparing to start an engine, first see that the nuts or studs holding cylinder head to cylinder are tight, as the heating and cooling of the cylinder are liable to loosen them. Then oil all bearings with a hand oil can, and carefully wipe off all outside grease.

When all is ready, work the gasoline pump to get the air out of the feed pipes and fill the reservoir.

First, the engine must be turned so that the piston is as far back as it will go, and to prevent air being pressed back the exhaust must be held open, or a cock in priming cup on top of cylinder opened.

If gasoline priming is needed, the gasoline must be poured into the priming cup after closing the cock into the cylinder, for it would do no good to merely let the gasoline run down into the cylinder in a cold stream: it must be sprayed in. If the exhaust has been held open, and the priming charge of gasoline is to be drawn in through the regular supply pipe and valve, the exhaust should be closed and the throttle turned on to a point indicated by the manufacturer of the engine.

We suppose that the igniter is ready to work. If the hot tube is used, the tube should be hot; if the electric igniter is used, the igniter bar should be in position to be snapped so as to close the circuit and cause a spark when the charge has been compressed.

If all is ready, open the cock from which the supply of gasoline is to be obtained, and at the same time turn the engine over so as to draw the charge into the cylinder. If a priming cock has been opened, that must be closed by hand as soon as the cylinder is filled and the piston ready to return for compression. If the regular feed is used, the automatic valve will close of itself.

Bring the flywheel over to back center so that piston will compress the charge. With the flywheel in the hand, bring the piston back sharply two or three times, compressing the charge. This repeated compression causes a little heat to be liberated, which warms up the cylinder inside. If the cylinder is very cold this compression may be repeated until the cylinder is sufficiently warm to ignite. When performing this preparatory compression the piston may be brought nearly up to the dead center but not quite. At last bring it over the dead center, and just as it passes over, snap the electric ignition bar. If an explosion follows the engine will be started.

If the hot tube is used, the flywheel may be brought around sharply each time so that the piston will pass the dead center, as an explosion will follow complete compression. If the explosion does not follow, the flywheel may be turned back again and brought up sharply past the dead center. Each successive compression will warm up the cylinder a little till at last an explosion will take place and the engine will be started.

More gasoline will be needed to start in cold weather than in warm, and the starting supply should be regulated accordingly. Moreover, when the engine gets to going, the cylinder will warm up, more of the gasoline will vaporize, and a smaller supply will be needed. Then the throttle can be turned so as to reduce the supply.

After the engine is started, the water jacket should be set in operation, and you should see that the cylinder lubrication is taking place as it ought.

As the above method of starting the engine will not always work well, especially in cold weather, what are called "self-starters" are used. They are variously arranged on different engines, but are constructed on the same general principle. This is, first, to pump air and gasoline into the cylinder instead of drawing it in by suction. Sometimes the gasoline is forced in by an air compression tank. The engine is turned just past the back center, care having been taken to make sure that the stroke is the regular explosion stroke. This may be told by looking at the valve cam or shaft. If an electric igniter is used, it is set ready to snap by hand. If the tube igniter is used, a detonator is arranged in the cylinder, to be charged by the head of a snapping parlor match which can be exploded by hand. Holding the flywheel with one hand with piston just past back center, fill the compressed end of the cylinder by working the pump or turning on the air in compression tank till you feel a strong pressure on the piston through the flywheel. Then snap igniter or detonator and the engine is off. If throttle valve has not been opened, it may now be immediately opened.

The skill comes in managing the flywheel with one hand, or one hand and a foot, and the igniter, etc., with the other hand. Care must be exercised not to get caught when the flywheel starts off. The foot must never be put through the arm of the wheel, the wheel merely being held when necessary by the ball of the big toe, so that if the flywheel should start suddenly it would merely slip off the toe without carrying the foot around or unbalancing the engineer. Until one gets used to it, it is better to have some one else manage the flywheel, while you look after the gasoline supply, igniter, etc. When used to it, one man can easily start any gasoline engine up to 15 horsepower.

WHAT TO DO WITH A GASOLINE ENGINE WHEN IT DOESN'T WORK.

Questions and Answers.

Q. If the engine suddenly stops, what would you do?

A. First, see that the gasoline feed is all right, plenty of gasoline in the tank, feed pipe filled, gasoline pump working, and then if valves are all in working order. Perhaps there may be dirt in the feed reservoir, or the pipe leading from it may be stopped up. If everything is right so far, examine the valves to see that they work freely and do not get stuck from lack of good oil, or from use of poor oil. Raise them a few times to see if they work freely. Carefully observe if the air valve is not tight in sleeve of gas valve.

Q. What would be the cause of the piston's sticking in the cylinder?

A. Either it was not properly lubricated, or it got too hot, the heat causing it to expand.

Q. Are boxes on a gasoline engine likely to get hot?

A. Yes, though not so likely as on a steam engine. They must be watched with the same care as they would be on a steam engine. If the engine stops, turn it by hand a few times to see that it works freely without sticking anywhere.

Q. Is the electric sparking device likely to get out of order?

A. Yes. You can always test it by loosening one wire at the cylinder and touching it to the other to see that a spark passes between them. If there is no spark, there is trouble with the battery.

Q. How should the batteries be connected up?

A. A wire should pass from carbon of No. 1 to copper of No. 2; from carbon of No. 2 to copper of No. 3, etc., always from copper to carbon, never from carbon to carbon or copper to copper. Wire from last carbon to spark coil and from coil to switch, and from switch to one of the connections on the engine. Wire from copper of No. 1 to the other connection on the engine. In wiring, always scrape the ends of the wire clean and bright where the connection is to be made with any other metal.

Q. What precautions can be taken to keep batteries in order?

A. The connections between the cells can be changed every few days, No. 1 being connected with No. 3, No. 3 with No. 5, etc., alternating them, but always making a single line of connection from one connection on cylinder to first copper, from the carbon of that cell to copper of next cell, and so on till the circuit to the cylinder is completed. When the engine is not in operation, always throw out the switch, to prevent possible short circuiting. If battery is feeble at first, fasten wires together for half an hour at engine till current gets well started.

Q. Is there likely to be trouble with the igniter inside cylinder?

A. There may be. You will probably find a plug that can be taken out so as to provide a peep hole. Never put your eye near this hole, for some gasoline may escape and when spark is made it will explode and put out your eye. Always keep the eye a foot away from the hole. Practice looking at the spark when you know it is all right and no gasoline is near, in order that you may get the right position at which to see the spark in case of trouble. In any case, always take pains to force out any possible gas before snapping igniter to see if the spark works all right.

Q. If there is no spark, what should be done?

A. Clean the platinum points. This may be done by throwing out switch and cutting a piece of pine one-eighth of an inch thick and one-half inch wide, and rubbing it between the points. It may be necessary to push cam out a trifle to compensate for wear.

Q. How can you look into peep hole without endangering eyesight?

A. By use of a mirror.

Q. If the hot tube fails to work, what may be done?

A. Conditions of atmosphere, pressure, etc., vary so much that the length of the tube cannot always be determined. If a tube of the usual length fails to work, try one a little longer or shorter, but not varying over $1\frac{1}{2}$ inches.

Q. When gas is used, what may interfere with gas supply?

A. Water in the gas pipes. This is always true of gas pipes not properly drained, especially in cold weather when condensation may take place. If water accumulates, tubes must be taken apart and blown out, and if necessary a drain cock can be put in at the lowest point.

Q. What trouble is likely to be had with the valves?

A. In time the seats will wear, and must be taken out and ground with flour or emery.

Q. Should the cylinder of a gasoline engine be kept as cool as it can be kept with running water?

A. No. It should be as hot as the hand can be borne upon it, or about 100 degrees. If it is kept cooler than this the gasoline will not gasefy well. If a tank is used, the circulation in the tank will justify the temperature properly. The water may be kept at 175 degrees of temperature, and used for hot water heating. The exhaust gases are also hot and may be used for heating by carrying in pipes coiled in a hot water heater.

Q. Are water joints likely to leak?

A. Yes. The great heating given the cylinder is liable to loosen the water joints. They are best packed with asbestos soaked in oil, sheets 1-16 inch thick. Old packing should always be thoroughly cleaned off when new packing is put in.

Q. How may the bearings be readjusted when worn?

A. Usually there are liners to adjust bearing. In crank box adjust as in steam engine by tightening the key.

Q. If you hear a loud explosion in the exhaust pipe after the regular explosion, should you be alarmed?

A. No. All gas or gasoline engines give them at times and they are harmless. If the gas or gasoline fed to the engine is not sufficient to make an explosive mixture, the engine will perhaps miss the explosion, and live gas will go into the exhaust pipe. After two or three of these have accumulated an explosion may take place and the burned gases coming out of the port as hot flames will explode the live gas previously exhausted. Any missing of the regular explosion by the engine, through trouble with battery, or the like, will cause the same condition.

Q. When you get exhaust pipe explosions, what should you do?

A. Turn on the fuel till the exhaust is smoky. Then you know you have fuel enough and more than enough. If the explosions still continue, conclude that the igniter spark is too weak, or does not take place.

Q. What precaution must be taken in cold weather?

A. The water must be carefully drained out of jacket.

Q. Will common steam engine cylinder oil do for a gasoline engine?

A. No. The heat is so great that only a special high grade mineral oil will do. Any oil containing animal fat will be worse than useless.

Q. How can you tell if right amount of gas or gasoline is being fed to engine to give highest power?

A. Turn on as much as possible without producing smoke. A smokeless mixture is better than one which causes smoke.

Q. If you have reason to suppose gas may be in the cylinder, should you try to start cylinder?

A. No. Empty the gas all out by turning the engine over a few times by hand, holding exhaust open if necessary.

Q. How long will a battery run without recharging?

A. The time varies. Usually not over three or four months.

Q. Is it objectionable to connect an electric bell with an engine battery?

A. Certainly. Never do it.

Q. If your engine doesn't run, how many things are likely to be the trouble?

A. Not more than four—compression, spark, gas supply, valves.

GAS AND GASOLINE ENGINES—Continued

Explanation of Principles.

Reference has already been made to the gas engine, and a general description is given of this interesting and useful machine, but as no detailed explanation is there given of the principles controlling its action, it was deemed wise by the author to place before his readers, additional matter pertaining to the subject, and in doing so an effort has been made to present it in as simple and plain a manner as possible, in order that it may be easily within the comprehension of all. As the gas, and gasoline engine are practically identical in principle, the same explanation and illustrations will, with the exception of a few minor details, apply to both.

All gas engines in practical use at the present time are two or four cycle, as herein described, or are modifications of these forms. Of these two types the four cycle machine has by far been the more generally adopted. We will now explain and illustrate the principles involved and the difference of action in the two types, taking up the four cycle first.

GAS ENGINES

Before proceeding farther, it will be well to explain the meaning of the word cycle as used in this connection.

A CYCLE means a round of time or a round of events necessary to produce a certain result.

As applied to the gas engine it means the round of movements or events to get one "explosion" as it is commonly called, or on impulse. In other words one working stroke.

In the four cycle engine we have four distinct movements or events to get one "explosion."

Beginning with Fig. 1 we show the piston C, starting on the first or downward stroke, drawing in, by suction, the charge or mixture of air and fuel through valve A.

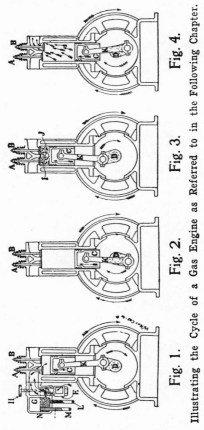

Fig. 4.   Fig. 3.   Fig. 2.   Fig. 1.

Illustrating the Cycle of a Gas Engine as Referred to in the Following Chapter.

This is the SUCTION or INTAKE stroke and is the first movement or event.

The valve A is now closed by its spring, and in Fig 2 we show the piston C returning, and compressing the charge into a small space, called the compression space, in the upper end of the cylinder. This is the COMPRES-SION stroke and is the second movement.

In Fig. 3 we show piston C almost to the top or end of the stroke. The mixture of air and gas is now under compression and at this point the electric spark is made which fires the charge. By the time the mixture is ignited, and the crank D reaches the center, the heat of the burning charge expands with great force and drives piston C down on its POWER stroke. This is the third movement.

In Fig. 4 we show piston C nearly to the end of the stroke. As much of the power of the heat expansion has been delivered as can be obtained, and at this point the exhaust valve B is forced open by a cam, and the remaining heat rushes out. Valve B is held open by the cam while piston C travels back to the top driving out the foul, spent gases. This is the SCAVENGING stroke and is the fourth and last movement in the cycle of operation. When the piston C, reaches the top, valve B closes and the engine is ready to begin the cycle or round of events over again.

We thus see that in a four cycle engine we have four movements—one cycle—giving it the short and well fitted name "four cycle."

Two revolutions of the flywheel are used to get the four movements of the piston. Many four cycle engines are made horizontal, that is, with the cylinder lying down instead of standing up as shown in the illustrations. The movements or actions, however, remain the same.

The two cycle engine draws in a charge, compresses it, fires it, and exhausts the burned gases, but it is all done in two strokes or movements, hence we call it a two cycle engine. In Figs. 5 and 6 we illustrate the action of the ordinary two cycle machine.

In Fig. 5 the air is drawn in at A, and fuel from B, as regulated by needle valve C. This mixture is drawn into the crankcase as the piston goes up, and a charge that is in the cylinder above the piston is compressed at the same time, thus drawing in a charge and compressing one, in one stroke or movement. When the piston reaches the top the compressed charge is fired, and the piston is driven down, delivering the power of the burning charge and also compressing the fresh charge just drawn into the crankcase. As the piston nears the end of the stroke, it uncovers the port E and the engine exhausts. An instant later the inlet port F is uncovered

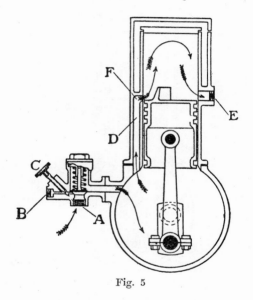

Fig. 5

as the piston moves on, and the fresh charge now under compression in the crankcase rushes up through the inlet port F, and is turned upward by the projection on the end of the piston.

As the new charge rushes in it is expected to drive out the burned gases at the exhaust port E. So we see the down stroke combines both the power and scavenging events, while the up stroke combines the intake and compression events. Two movements do here what four

are required to do in the four cycle engine. This type
is known as the two port two cycle engine.

   In Fig. 6 air and fuel are drawn in at port A by the
vacuum the piston produces, in the crankcase, on its
upward stroke.   By this construction the check valve,
shown in Fig. 5, is not needed. Otherwise the action
illustrated in Fig. 6 is the same as in Fig. 5. The type
shown in Fig. 6 is known as the three port two cycle
engine.

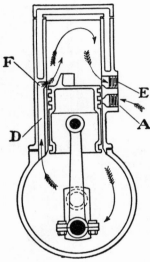

Fig. 6

   There are many other forms for the mechanical construction of the two-cycle engine, but they all follow the
same general principles we have here illustrated and
described.

   Owing to its simple construction and an impulse every
two strokes or every revolution, the two cycle engine
has proven very attractive to hundreds of inventors, and
a great variety of designs have been built. Simplicity of
construction is much to be desired in any machine if it

produces the results we want. The results demanded of a gas engine are economy in the use of fuel and re- liability of action.

The four cycle engine, in spite of its valves and gears, has given so much better practical results as a rule, that it has been adopted by most of the builders. The two cycle engine finds its best applications in service where the load and speed are comparatively uniform.

Comparing the machines we have illustrated, we see a charge drawn in every revolution by the two cycle construction. Part of the charge may be lost, by leak- age through the bearings, when the piston comes down compressing it in the crankcase.

The charge next passes to the cylinder and as the piston returns to the top it compresses the charge a second time, a loss of net power to the engine.

As the two cycle engine, when working properly, takes a charge and burns one every revolution, it would seem, at first thought, that it should give twice as much power as a four cycle machine of the same cylinder di- mensions. Owing to the losses we have mentioned, and the fact that the scavenging (driving out of the burned gases) is uncertain, the two cycle develops only 20 per cent to 50 per cent more power than the same size four cycle cylinder. As the bearings wear, the loss from the crankcase is apt to become greater, and the port action also changes slightly.

One construction of the two cycle engine avoids crank- case compression, and the losses thereby, by a design similar to steam engine practice. The engine is made with a crosshead and piston rod. The end of the cylin- der, usually left open by gas engine builders, is closed and fitted with a stuffing box through which the piston rod works exactly like the steam engine. This closed end of the cylinder is used, instead of the crankcase, for handling the mixture or charge.

The principles involved in the two cycle engine de- pend largely on a certain velocity for the moving or transferring of the charge from the receiving and first compression chamber to the cylinder; hence as wide a variation in the speed cannot be permitted as with the four cycle machine.

FUELS FOR THE GAS ENGINE.

Gas in its natural form, as found in some places, is the most convenient fuel known for the gas engine, especially for stationary work.

Only a few sections, however, are so favored, and in other places, and for portable or traction work the gas for the engine must be made or produced artificially from the most available substance.

At the present time gasoline is used more than anything else owing to the ease with which it is carburetted, or converted into gas just as it is needed by the engine.

Natural gas and gasoline have been used so much more than other fuels that the expression "gas and gasoline engines" as frequently used is taken by some to mean two distinct types of engines.

The expression as heretofore explained is misleading, for the general principles are precisely the same. The only difference in construction is in the compression used, and in the mixer or device for feeding the fuel; it is a simple matter to change a gas engine to gasoline, or a gasoline engine to gas.

When we say "gas engine" we have covered the ground, and we understand that in using gasoline, oil, coal, etc., proper means must be provided for converting the fuel used into gas.

Many fuels will produce gas which may be used in the gas engine. Among these are coal, crude oil, coal oil or kerosene, gasoline, wood alcohol, spirits of various kinds, etc.

A great many of the possible fuels are out of the question because of price, and others involve difficulties in the way of generating or producing the gas as needed and of proper quality. Some gases also involve objectionable features in the burning or combustion, as for example, the gas from crude oil (a very cheap fuel) carries with it a carbon element that is deposited on the head of the piston and on the walls of the compression space, making it necessary to clean these parts frequently.

The difficulties in generating or producing gases, and in burning them to produce power, are being rapidly overcome by new improvements and methods, and the advantages of the gas engine are increasing thereby.

Gas from coal is proving to be a very cheap fuel for gas engines for heavy and stationary work.

One ton of coal used in this way, as proved by actual practical results, will do two to three times the work that it will do by burning it under a steam boiler.

As this book has reference, more especially, to gas engines for light portable and traction work, we will pay particular attention to gasoline as the available fuel possessing the most advantages.

Carbureters or Mixers.—Several different types of mixers or carbureters for gasoline are in common use, the principle we illustrate in Fig. 1, probably being most generally used.

G is an overflow chamber holding the gasoline at a certain level in standpipe F, as indicated by the dotted line N. Gasoline flows into the chamber, G, from a pump, through pipe, L, and the overflow goes back to the tank through pipe M.

As the air is drawn into the cylinder, through the air regulator E, it pulls gasoline with it from standpipe F, the amount being regulated by the needle valve H.

This may be called the constant level overflow system, and is generally built in as part of the engine proper, in the plain, heavy engines now common for stationary work.

If a float was placed in chamber G, operating a gasoline inlet valve, and the gasoline tank was placed higher than the chamber, we would then have the float feed carbureter system. The float would hold the gasoline at a certain level in the standpipe just as the overflow in Fig. 1.

The float feed carbureter is generally a separate part or adjunct to the engine, and it is common for this part to be made by the manufacturer of parts or specialties.

Figure 7 is a cross-section of a float feed carbureter. The float M controls the valve O, and holds the gasoline at a certain level in the spray nozzle L. The air supply in starting, or at slow speed of the motor comes through the narrow passage I, and in passing the spray nozzle L, it draws a small quantity of gasoline as regulated positively by needle valve A. J is the connection to the engine and K is a throttle to enable the operator to control

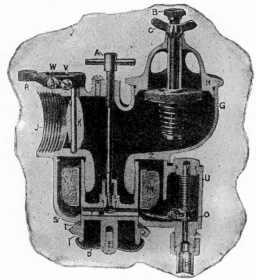

Fig. 7

the quantity of mixture admitted to the cylinder. The air valve F, is held to its seat by a light spring G, with tension adjusted by screw B and locking device C.

As the throttle K is opened, admitting more mixture to the engine, the air valve A opens wider, admitting more air. As the suction on the gasoline spray nozzle L is greater, more gasoline is drawn, thus keeping the proportionate mixture approximately right under the throttle, and at the various engine speeds. Both air and gasoline are thus automatically measured, under the varying conditions.

Figure 8 shows a float carbureter with a different principle.

A is the connection; B, gasoline needle valve; C, constant air inlet; D, compensating, or automatic air valve, with spring tension regulated by E; F is the gasoline

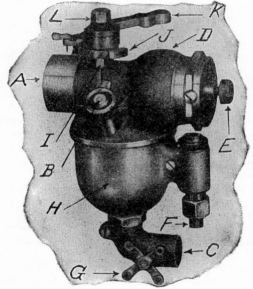

Fig. 8

pipe connection; G is a throttle in the air passage C; H, float chamber; I, needle valve control lever; J, cam, operating mixture throttle lever; L, nut for adjusting lever to position desired.

In starting, the constant air passage C, is partly close 1, to secure more suction on the gasoline. As the throttle, operated by lever K is opened, the cam J moves lever I, gradually opening gasoline needle valve B, admitting more gasoline in proportion to the increased air supply coming through the air valve D.

In this carbureter the air supply is automatically controlled, but the gasoline is positively regulated.

It is evident that the gasoline tank must be higher than the carbureter to supply gasoline to the flat valve by gravity.

The gasoline may be supplied from a lower level by air pressure in the gasoline tank, but as this is complicating the mechanism of the outfit, it is rarely used.

It will be observed that the general principles of the overflow and feed float systems are the same.

Another very common method of feeding gasoline is by means of a mixing valve as illustrated in Fig. 5.

The gasoline, regulated by the needle valve C, feeds to the seat of the mixing valve. When the piston draws in air from A, the mixing valve is lifted, gasoline flows in and is mixed with the air.

At the end of the suction stroke the valve closes, shutting off the gasoline. The gasoline may be supplied to the valve by gravity or by air pressure, the same as with the float feed system.

A fault with the ordinary mixing valve is found in the fact that, as the valve closes, the fuel remaining on the seat is projected backward by the angle of the seat, causing the valve to "spit" or "slobber" the fuel.

In all these systems of feeding, the gasoline is so volatile that, by mixing with the air as it is drawn in and in passing into the heated cylinder, it is carburetted or vaporized, so that by the time the spark is made, the mixture is formed and ready to be ignited.

Fuel gas for gas engines is made from the heavy crude oils by subjecting the oil to heat in a special producer apparatus, that makes a gas vapor from the volatile parts of the oil, while separating and retaining the heavy, solid matter.

Too much gas in the cylinder will not burn for want of sufficient air, just the same as a furnace fire will not burn if the dampers are closed.

Too much fuel turned on in starting is a frequent

cause of a gas engine refusing to start. In this case close the needle valve, and turn the engine until the surplus fuel has been driven out.

No matter what kind of fuel gas is used, the principle of feeding to the engine remains the same—A CERTAIN QUANTITY OF GAS WITH THE RIGHT AMOUNT OF AIR, must be taken for each "explosion" or impulse.

It must also be remembered that solid and liquid fuels *must be converted into gas* before the engine will run.

In using gasoline the natural heat of the air is generally depended upon for vaporizing, or making enough gas to start the engine. In the winter season the air frequently does not possess the required warmth, or heat for starting, often causing trouble to the inexperienced operator. In this case the necessary heat to supply the first charges of gas must be provided.

After the first few "explosions" there will be enough heat in the engine cylinder to vaporize the gasoline in the coldest weather.

Gases vary a great deal in the heat power or heat units possessed, and for this reason different gases will increase, or decrease the power of an engine of given size.

Gas from gasoline is very powerful, furnishing another excellent reason for its common use with gas engines.

Compression.—Compressing the charge or mixture of air and gas, before igniting it increases the force of the "explosion."

The higher the compression can be successfully carried, the greater will be the power derived from a given amount of fuel.

The compression heats the mixture rapidly, and, if the compression is carried too high, this heat will fire the charge before time for the spark, and before the piston reaches the end of the stroke. This would cause some of the force to be applied in the backward direction, and cause the engine to "pound" or perhaps stop.

The amount of compression that can be successfully used, depends on, first, the kind of fuel gas that is to be used; second, the speed for which the engine is designed, and third, the uniform heat of the cylinder walls and head at all times.

Different gases require higher or lower compression to obtain the best results, as for example, natural gas will admit of much higher compression than the gas from gasoline.

An engine built for high speed will carry a higher compression than could be used at low speed. The piston coming up to the end of the stroke so much faster enables the crank to pass the center before the impulse begins, even though the charge should be self-ignited from the heat of the high compression. Reliable, even temperature of the cylinder walls and head is of great importance for a high compression, for if the walls become overheated at times, the compression heat will be greater; if the compression is already up to the limit, this extra heat will cause pre-ignition, or firing of the charge too soon.

While it is desirable, from the standpoint of economy in fuel, and maximum or greatest power for a given cylinder dimension, to use the highest compression possible, yet no rule can be given that will fit different makes of engines for the reasons given above.

The manufacturer must be depended upon for the highest compression practical in his particular engine, to suit the design, speed, and fuel to be used.

As we are referring to gasoline as the most convenient and practical fuel for light portable, and traction work, we might say that a fair average compression for this fuel would be 60 lbs. gauge pressure, but the reader will understand that it may be more or less depending on the conditions as stated. This would be equal to about five atmospheres—that is, the volume of the cylinder and compression space would be squeezed up into a space one-fifth the total volume.

This amount of compression will, under proper conditions, give about 300 lbs. per square inch, heat expansive force, or "explosive" pressure at the moment of complete ignition.

A compression of 40 lbs. gauge pressure will give an initial "explosive" force of about 225 lbs. per square inch, so we see, as stated, that the net working force increases as we increase the compression.

The gauge pressure of the compression may be determined by the method described under the heading "How to Test the Condition of an Engine."

If a different gas fuel, from that for which the engine was sold, is to be used, it would be advisable to write the manufacturer of the engine as to the proper compression, as shown by factory tests, and how to change the compression space to best advantage. This will save much time and trouble in experimenting to obtain the best possible results.

The efficiency and economy of a gas engine depends in large measure on perfect compression, and any leakages in rings, valves, packings or porous cylinder walls, directly affect the working of the machine.

The building of an engine for the highest possible compression is a matter of close and careful study for the designer only, hence we will not go into details of construction.

In the operator's hands any make of engine must be carefully guarded against leakages of compression, if the highest possible efficiency and fuel economy are desired.

Ignition Apparatus.—In the early stages of the development of the gas engine, the charge of air and gas was ignited by a hot tube. This tube, with the outer end closed, was screwed into position on the engine and connected with the compression space. The tube was enclosed by a casing lined with heavy asbestos, and was kept at an intense heat by a gas fire within the casing. A part of the charge or mixture was forced into the tube by the compression stroke when it would be ignited by the fierce heat of the outer closed end.

This system, clumsy and crude in the light of late improvements, is known as hot tube ignition. It required time in starting to properly heat the tube; it was wasteful in the use of fuel, and the fire to heat the tube was a source of danger. Waste of fuel was due to maintaining the fire to heat the tube, and to the fact that the time of ignition was not under perfect control. Tubes burning out frequently added to the troubles.

The ignition or firing of the charge by an electric spark, under control at all times, is one of the great improvements in the gas engine, and has had much to do with bringing the machine into favor with power users.

The spark is made on the inside of the cylinder, thus eliminating the danger of fire with the hot tube. By this improvement the gas engine became a safe power generating machine in the strictest sense of the word.

Electric ignition has come into such general use that the hot tube is now seldom made, unless for emergency use and most manufacturers do not furnish it at all.

There are two systems of electric ignition in general use, viz: the primary, or make and break, and the secondary or jump spark. Both of these systems must have a source of electric current; a coil for storing, and discharging the current, a device for making and breaking the circuit, and an igniter or spark plug as the case may be.

The source of electric current for either system may be a battery, or a generator driven by the engine. Where a generator is used it is generally considered necessary to have batteries for starting, and switch over to the generator after the engine gets up speed. Most generators require more speed, to furnish the necessary current, than the operator would be able, or willing to give it in starting the engine.

The spark coil acts as a sort of reservoir to store up current when the circuit is made, and to discharge it when the circuit is broken, and this discharge between two points, inside the compression space, makes the spark that fires the charge.

With the make and break system, the circuit is made and broken inside the compression space, giving this system its name.

The contact, or make and break points are set in a block or carrier, the whole forming a device known as the igniter.

The contact points are called electrodes, one of which

is made stationary and insulated by a non-conducting material from the other parts of the engine. The other electrode is movable, and the mechanism of the engine causes it to form a contact, inside the compression space, with the insulated stationary electrode, just an instant before time for the spark. During this very short time of contact the current from the battery or generator flows through and charges the coil. At the right moment the movable electrode is snapped back, breaking the contact with the insulated electrode, and the current, stored in the coil, is discharged across the gap between the contact points or electrodes, causing the spark. The quicker the break is made the better and stronger will be the spark produced.

The spark coil for primary or make and break ignition consists of a bundle or core of soft iron wire around which is wound a quantity of insulated copper wire called a primary winding. The current from this coil is a primary current, which explains why make and break ignition is also called primary ignition.

For the secondary or jump spark system the spark coil receives another winding, of several thousand feet of fine insulated wire, called the secondary winding.

This makes a jump spark, or high tension coil as the secondary winding carries a current of high voltage. The electric circuit for this system of ignition is interrupted at any suitable, convenient point on the engine, and causes a spark to jump between two stationary points inside the combustion chamber.

These two points are carried by a spark plug that is screwed into an opening to the combustion chamber, and one of the points must be insulated so the current will pass around and jump the gap provided. The device for interrupting the circuit in the jump spark system is a timer, sometimes called the "commutator" and is shown in Fig. 9 at D. The break of the contact points of the timer must be very quick, and produces a spark at the gap between the points of the spark plug.

It has become quite common to provide the spark coil with an automatic vibrator; the instant the timer makes the circuit, the automatic vibrator sets up a vibrating

motion producing a string of sparks at the plug instead
of one. With the automatic vibrator, the very quick part-
ing of the timer contact is not essential.

The merits of the vibrator as against the plain jump
spark coil have been much discussed. Some authorities
claim the plain coil is less liable to get out of
order, not having a delicate vibrator adjustment; that
making one good spark does the work, which is all that
is required.

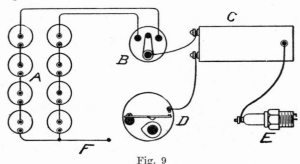

Fig. 9
Wiring Diagram—Jump Spark Ignition.
A—Batteries.
B—Switch.
C—Jump Spark Coil.
D—Timer or "Commutator."
E—Spark Plug.
F—Ground Wire to Engine.

It is claimed for the vibrator coil that a more sure
and rapid ignition is obtained; that the delicate adjustment of the vibrator is a simple matter, and not a dis-
advantage in the hands of the intelligent operator, and
that the necessary quick make and break of the circuit,
being made automatically, insures perfect ignition at
any speed of the engine. As the vibrator coil has come
into general use it must be conceded that the majority
of users think it has advantages which overbalance its
disadvantages.

The illustration, Fig. 9, shows how to connect two
sets of batteries to a switch so one set may be used while

the other is out of circuit—thus holding an extra battery in readiness for immediate service should the set in use fail.

When the switch B, is in the central position as shown, both batteries are out of circuit.

In connecting up a jump spark ignition outfit it will generally be found that the manufacturer of the coil has marked the terminals or binding posts. "Bat." stamped on the coil means to attach the battery to that binding post. "Com." means the connection to the timer or "commutator," while "Sec." denotes the terminal of the secondary winding, to be connected to the spark plug.

Should there be two secondary terminals on the coil, one may be connected to the terminal marked "com." This is usually done within the coil, leaving but three outside connections as shown in Fig. 9.

The switch is placed between the battery and the coil and it is understood that the "Bat." connection on the coil is carried to the switch and then on to the battery.

A generator, made for jump spark ignition may be connected to the switch instead of one of the batteries, similar to the connections for primary ignition, illustrated in Fig. 10, which shows the wiring for a make and break ignition outfit, using a battery for starting and connections to switch the generator into circuit as soon as the engine gets up the speed necessary to make the generator deliver the required current.

Some generators are advertised as furnishing current at a very low speed and thereby dispensing with the battery. Most generators, however, will require more speed than the operator would be able or willing to give it in starting the engine.

It is important, in connecting up an ignition outfit, to see that the wire terminals, and binding posts are clean and that the connections are made secure. The ground wires may be attached at any convenient point on the engine, but paint and grease must be removed to secure a good circuit.

Flexible wiring is less liable to break at the point of connection and cause trouble. A solid copper wire will frequently break close to the binding post and remain in position apparently sound.

Spark coils are manufactured by specialists and their manufacture, on a large scale, has been so well developed that the selling price is too low for the engine manufacturer to think of making his own coils.

The wire, from the secondary winding of the jump spark coil to the spark plug, should be heavily insulated, or care must be taken to keep it clear of other parts that would complete a circuit, owing to the high voltage current that would leak and short circuit through a light in-

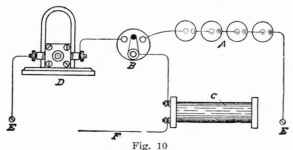

Fig. 10

Wiring Diagram—Battery and Generator. Make and Break Ignition.

A—Battery.
B—Switch.
C—Simple Primary Coil.
D—Magneto or Generator.
EE—Ground Wires to Engine.
F—Wires to Stationary Electrode of Igniter.

sulation. This is a frequent cause of trouble with jump spark ignition. A light or faulty insulation on the secondary wire will often deceive the inexperienced operator.

A bare wire, from the secondary binding post on the coil to the plug, would be better than a defective insulation, for with the bare wire every one would know it must not touch other objects that would conduct the current.

There is a great deal of discussion among gas engine builders and users concerning the relative merits of the two systems of electric ignition, but as both give good service and satisfaction under proper conditions, it is a matter that can be decided only in individual cases. It is argued for the make and break system that the low voltage current is less liable to short circuit; that the coil is much cheaper, and less liable to go wrong, and that a bigger and better spark is made.

For the jump spark ignition a great advantage is claimed by doing away with the movable electrode, its wear and consequent leakage from the combustion chamber; that the time of the spark can be easily retarded and advanced at will to suit all speeds and conditions, and that if the coil and high tension current are handled intelligently they will not fail, but will go on indefinitely doing their work faithfully.

It is generally considered that jump spark ignition is better suited for high speed engines, while the low speed, heavily constructed engines commonly used for stationary work are usually equipped with the make and break system. So far as practical application is concerned either system can be applied to suit any condition.

As both are good, the reader will be left to decide for himself as his own experience or preference may direct.

Timing the Valves and Spark—The valves of the gas engine are almost universally of the poppet variety, and are operated by cams and springs which produce a very quick opening and closing action. In order to obtain a high efficiency in the working of the engine it is necessary that the valves open, and that the spark occur at the proper moment, to produce the best results. The inlet valve A, shown in Fig. 1, is of the automatic type, being opened by the suction stroke of the piston. While many gas engines are built this way, it is quite common to open the inlet, as well as the exhaust valve mechanically, or by means of a cam operated by the engine.

The automatic inlet valve, as its name implies, is self timed, opening at the beginning, and closing at the end of the suction stroke.

The cams to open all mechanically operated valves must be set or timed with reference to the position of the crank and piston. The exhaust valve should be opened about 40° before the crank on its power stroke reaches center.

In an engine with 6″ stroke, the piston would be about $1\frac{1}{8}$″ from the end of the stroke.

This last part of the stroke is not effective in delivering power to the crank shaft, and the exhaust valve is opened thus early to get rid of the remaining heat as soon as it becomes useless and thus have the cylinder in better shape to receive the next charge. The exhaust valve should not close before the end of the scavenging stroke, and not later than 20° past dead center.

If the inlet valve is operated mechanically, the cam should be set to open and close the valve when the crank has passed the dead centers 10°, to 20° according to the speed of the engine.

The late closing of the inlet valve on high speed engines is to allow the inertia or moving force of the incoming charge to increase the power of the cylinder by increasing the amount or volume of mixture taken in. Some claim the crank may pass the center 30° to 40° before the mixture stops coming in, although the piston has traveled back on the compression stroke one-half inch or more. The possible advantage is a slight increase of power from a cylinder of given size.

The timing of the ignition is of much greater importance than was realized for many years after the gas engine came into use. Although a proper mixture under compression fires easily and burns rapidly, yet it requires a little space of time, and the spark must occur far enough ahead of the center so the charge will be aflame, and the expansion taking place when the piston and crank start on the power stroke. If the spark comes too late, a part of the effective impulse stroke is lost, while if the spark is made too early, the heat expansion begins before the crank reaches the center and some of the power is thus delivered in a backward direction. This will cause the engine to "pound" or perhaps stop, if the ignition is very much too early.

The correct time for the spark depends on the fuel

used, and speed of the engine. At high speeds the spark must be advanced or made further ahead of the center to give the necessary time for ignition, while at low speeds the spark must be retarded or made later.

It is necessary to provide high speed engines with a device for retarding the spark in starting, and changing to the advanced position after the engine gets up speed.

For very high speeds the spark must be produced somewhere from 60° to 90° ahead of center and this position, with the slow speed in starting, would deliver all the power in a backward direction, causing the engine to "kick."

Owing to the greatly varying speeds used it is impossible to give a set rule for the correct point of ignition, but the proper timing of the spark can be readily determined by a little experimenting. The operator will soon learn the correct position by observing the results of early or late ignition.

It is needless to say, that if the spark is too far advanced in starting the operator will soon find it out, for the engine is sure to make a "kick" about it.

A gas engine will run with the valves and spark considerably out of time, but its full power and efficiency will not be developed unless the timing is right.

As the inlet and exhaust valves, in proper turn, only open every second revolution of the crank shaft (with the four-cycle engine), the reader will understand that the cams are located on the back geared shaft, which runs at just one-half the speed of the crankshaft.

In timing the valves the question naturally arises when is the crank exactly at the end of the stroke or on "dead center?"

The crank travels a considerable distance at each end of the stroke, with but little perceptible movement of the piston and this fact gives considerable range in setting the valves while not greatly affecting the results.

Some users, especially of small engines, guess at the center by noting the piston's movement, but for the benefit of readers who insist on *knowing* when the crank is at center, we illustrate in Fig. 11 the following method:

With the crank turned to one side of center, as shown, insert a rod A, through an opening in the head of the engine allowing the rod to rest against the piston. Mark on the rod at B to show the distance to the piston and also mark the balance wheel at a fixed, stationary pointer C provided on the engine. Now turn the engine until the crank is on the other side of center as shown by the dotted lines. This position is determined by bring-

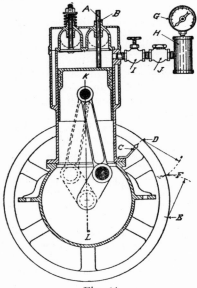

Fig. 11

ing the piston to the same distance from point B as shown by the mark we placed on the rod A.

Now make another mark on the balance wheel at the stationary pointer C. The two marks D and E on the balance wheel are at equal distances from the central position for the crank, and it follows that in bisecting or equally dividing the distance between the marks D and E and turning the engine so the central mark F,

comes to the stationary pointer C, we have thus brought
the crank to the "dead" center.

The opposite center is determined in a like manner.
The crank is thus brought at each end of the piston
stroke exactly to the center line K-L. Having estab-
lished the center we can readily calculate the degrees
from this for the opening and closing of the valves.
The circumference of the wheel is always equal to
360 degrees. If we divide 360 by the circumference
in inches we will know how many degrees in each inch.
To find a point 40 degrees from "dead" center divide
40 degrees by the number of degrees in an inch of the
circumference. The result will be the number of inches
from center to the point desired.

In the illustration, Fig. 11, it will be noticed that one
of the valves has been removed to insert the rod A
through the valve stem guide. By using a rod that fits
the guide the two positions of the piston, at equal
distance from the center can be accurately determined.
For engines that do not have the valves in the head
any other openings, such as for the spark plug or an
igniter, may be used, but it would be advisable to use
a special plug, or plate to fit the opening through which
a hole, to fit rod A, may be drilled.

The above method for locating dead center is the
same that is generally used for the steam engine except
that the mark B on the rod A is made on the crosshead
and guides.

As the gas engine ordinarily has no crosshead the process we have described and illustrated will be found
equally effective and simple, while, as with the steam
engine, it is mechanically correct.

TESTING THE CONDITION OF A GAS ENGINE.

First see that the valves are correctly timed.

The next thing to know is that the fuel reaches the
mixing chamber or carbureter. Now look after the compression to see if there is any serious leakage through the
rings, valves or packed joints.

Oil the engine thoroughly using care to know that the
cylinder walls are well lubricated with good gas engine

oil, then as a quick, ordinary compression test, sufficient for practical purposes, the engine is revolved bringing the piston up quickly on the compression stroke and holding it at the highest point of compression to see how soon the pressure will disappear. This may properly be called "feeling of the compression" and after a little experience the operator will be able to judge pretty accurately as to what results may expected of the engine.

The only recourse when serious leakage through the rings occurs, is new rings, or perhaps re-boring of the cylinder, new piston and new rings.

This is a job for the machinist.

After knowing that the fuel gets to the engine in proper time, and that the compression is all right, next look after the ignition apparatus, a very important part of every gas power machine.

The make and break system may be tested out as follows: Throw in the switch, then detach the wire from the stationary, insulated electrode of the igniter, and scrape it on the binding post from which it was removed. If a spark is produced with the igniter contact points open it will prove the insulation of the stationary electrode to be faulty. Should no spark appear, next close the contact points and scrape the wire again on the binding post. A good spark should now be produced. If not, go over the wiring very carefully to see if all connections are clean and secure, and to look for possible leakage of all the current, or short circuiting as it is commonly called.

Next remove the igniter to see if the contact points are corroded thus preventing the passage of the current.

While having the igniter detached it is a good plan to hold it to the engine and snap or break the contact points apart as when the engine is running. If, with the contact points clean, connections all properly made and no short circuits, a spark is not yet obtained, next look after the source of current (battery or generator as the case may be), and the trouble will soon be located in an exhausted battery, or in case of a generator it may be bad brushes or possible loss of speed if the generator is driven by belt or friction.

Once in a great while the spark coil may fail, but this is a rare occurrence, if the ignition apparatus is kept in a dry place as it should be.

Briefly stated, see that the engine gets the fuel in the proper time; see that there is no serious leakage and see that a good spark is produced at the right time.

These things in proper order and assuming, of course, in case liquid fuel is used, that the proper condition is present for carburetion, or vaporization, the engine is ready to run and may be depended upon.

The routine for testing a jump spark ignition outfit is similar to that just described for the primary or make and break system. By detaching the spark plug and allowing it to rest on the engine, so the circuit will be the same as when the plug is in position, work the circuit interrupting device (if a plain jump spark coil is used), or in case of a vibrator coil, turn the engine until the circuit is made by the timer, when the vibrator, if properly adjusted, will set up the buzzing sound familiar to users of vibrator coils.

A good spark should now appear between the points of the spark plug. If not, detach the wire from the plug, and holding the end of the wire within one-sixteenth to one-eighth inch of some part of the engine again work the trembler or make contact with the timer.

If a spark can now be produced it proves the insulation of the plug faulty, while should no spark appear next look for bad connections, short circuit or further back to the source of current as with the make and break system.

Now, knowing that the engine takes the charge and fires it properly, next see that the cooling system is in working order. If the cooling jacket, or passages formed in the castings of the cylinder and head for allowing the cooling element, oil, water or other liquid to circulate, should become clogged or choked the heat of the cylinder will rise too high, so in testing the condition of a machine we must examine the cooling facilities, and know that sufficient radiation of excess heat is maintained. This means, of course, that the proper circulation of the heat carrying agent (whether it be the water, oil, or air) must be provided.

The compression test described in the fore-going is a quick, offhand way of sizing up the running condition of small and medium sized engines, but it can only give an approximate idea of the amount of the compression. A very good method of obtaining the gauge pressure of the compression is illustrated in Fig. 11.

A pressure gauge G, is attached to a receiving chamber H, which is connected to the compression chamber of the engine by a globe valve I, and check valve J.

Run the engine up to full speed, then throw out the switch and immediately open the valve I. The highest compression pressure will be accumulated in the chamber H, and the gauge will register the pounds per square inch.

The valve I must not be opened while the engine is yet firing the charges, but it should be opened very quickly after the firing has stopped so the compression pressure may be registered at practically the normal running speed of the engine.

This test of the compression is not necessary to the successful care and operation of a gas engine for the manufacturer of the machine has, of course, figured out the best compression for the kind of fuel to be used and the work to be done.

We describe and illustrate the gauge test for the benefit of readers who may wish to make a deeper study of the gas engine, and gas engine principles than is necessary for the ordinary user.

THE SCIENCE OF
THRESHING

CHAPTER XIV.

HOW TO RUN A THRESHING MACHINE.

A threshing machine, though large, is a comparatively simple machine, consisting of a cylinder with teeth working into other teeth which are usually concaved (this primary part really separates the grain from the husk), and rotary fan and sieves to separate grain from chaff, and some sort of stacker to carry off the straw. The common stacker merely carries off the straw by some endless arrangement of slats working in a long box; while the so-called "wind stacker" is a pneumatic device for blowing the straw through a large pipe. It has the advantage of keeping the straw under more perfect control than the common stacker. The separation of the grain from the straw is variously effected by different manufacturers, there being three general types, called apron, vibrating, and agitating.

The following list of parts packed inside the J. I. Case separator (of the agitative type) when it is shipped will be useful for reference in connection with any type of separator:

2 Hopper arms, Right and Left,
1 Hopper bottom,
1 Hopper rod with thumb nut,
2 Feed tables,
2 Feed table legs,
2 Band cutter stands and bolts,
1 Large crank shaft,
1 Grain auger with 1223 T. pulley and 1154 T., Box,

1 Tailings auger,
1 Elevator spout,
1 Elevator shake arm, complete,
1 Set fish-backs, for straw-rack,
1 Elevator pulley, 529 T.,
1 Beater pulley, 6-inch 1254 T., or 4-inch 1255 T.,
1 Elevator drive pulley 1673 T.,
1 Crank pulley to drive grain auger 1605 T.,

1 Cylinder pulley to drive crank 4-inch 973 T., or 6-inch 1085 T.,
1 Cylinder pulley to drive fan 1347 T., 1348 T., or 1633 T.,
1 Fan pulley, 1244 T., or 1231 T.,
1 Belt tightener, complete, with pulley,
1 Belt reel, 5016 T., or 1642 T., with crank and bolt,
4 Shoe sieves,
4 Shoe rods, with nuts and washers,
1 Conveyor extension,
1 Sheet iron tail board,
2 Tail board castings 1654 T., and 1655 T.

In addition to these are the parts of the stacker.

As each manufacturer furnishes all needed directions for putting the parts together, we will suppose the separator is in working condition.

A new machine should be set up and run for a couple of hours before attempting to thresh any grain. The oil boxes should be carefully cleaned, and all dirt, cinders, and paint removed from the oil holes. The grease cups on cylinder, beater and crank boxes should be screwed down after being filled with hard oil, moderately thin oil being used for other parts of the machine. Before putting on the belts, turn the machine by hand a few times to see that no parts are loose. Look into the machine on straw rack and conveyor.

First connect up belt with engine and run the cylinder only for a time. Screw down the grease cup lugs when necessary, and see that no boxes heat. Take off the tightener pulley, clean out oil chambers and thoroughly oil the spindle. Then oil each separate bearing in turn, seeing that oil hole is clean, and that pulley or journal works freely. The successive belts may then be put on one at a time, until the stacker belt is put on after its pulleys have been oiled. Especially note which belts are to run crossed—usually the main belt and the stacker belt. You can tell by noting which way the machinery must run to keep the straw moving in the proper direction.

Oiling on the first run of a machine is especially important, as the bearings are a trifle rough and more liable to heat than after machine has been used for some time

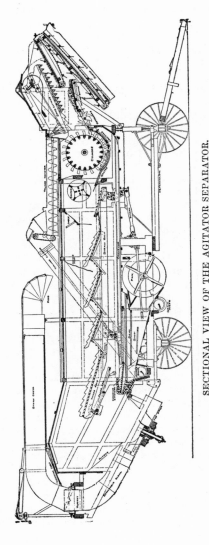

SECTIONAL VIEW OF THE AGITATOR SEPARATOR.

It is well to oil a shaft while it runs, since the motion helps the oil to work in over the whole surface.

The sieves, concaves, check board and blinds must be adjusted to the kind of grain to be threshed. When they have been so adjusted the machine is ready to thresh.

SETTING SEPARATOR.

It is important that the machine be kept perfectly steady, and that it be level from side to side, though its being a little higher or lower at one end or the other may not matter much. If the level sidewise is not perfect the grain will have a tendency to work over to one side. A spirit level should be used.

One or more of the wheels should be set in holes, according to the unevenness of the ground, and the rear wheels should be well blocked. Get the holes ready, judging as well as possible what will give a true level and a convenient position. Haul the machine into position and see that it is all right before uncoupling the engine. If holes need redigging to secure proper level, machine may be pulled out and backed in again by the engine. When machine is high in front it can easily be leveled when engine or team have been removed, by cramping the front wheels and digging in front of one and behind the other, then pulling the tongue around square.

Block the right hind wheel to prevent the belt drawing machine forward. Always carry a suitable block to have one handy.

In starting out of holes or on soft ground, cramp the front axle around, and it will require only half the power to start that would be required by a straight pull.

In setting the machine, if the position can be chosen, choose one in which the straw will move in the general direction of the wind, but a little quartering, so that dust and smoke from engine will be carried away from the men and the straw stack. In this position there is less danger from fire when wood is used.

THE CYLINDER.

The cylinder is arranged with several rows of teeth working into stationary teeth in what is called the concave. It is important that all these teeth be kept tight, and that the cylinder should not work from side to side. The teeth are liable to get loose in a new machine, and should be tightened up frequently. A little brine on each nut will cause it to rust slightly and help to hold it in place. If the cylinder slips endwise even a sixteenth of an inch, the teeth will be so much nearer the concaves on one side and so much farther away from them on the other side. Where they are close, they will crack the grain; where they are wide apart they will let the straw go through without threshing or taking out the grain. So it is important that the cylinder and its teeth run true and steady. If the teeth get bent in any way, they must be straightened.

The speed of the cylinder is important, since its pulley gives motion to the other parts of the machine, and this movement must be up to a certain point to do the work well. A usual speed for the cylinder pulley is 1,075 revolutions per minute, up to 1,150.

There is always an arrangement for adjusting the cylinder endwise, so that teeth will come in the middle. This should be adjusted carefully when necessary. The end play to avoid heating may be about 1-64 of an inch. It may be remembered that the cylinder teeth carry the straw to the concaves, and the concaves do the threshing.

THE CONCAVES.

The concaves are to be adjusted to suit the kind of grain threshed. When desiring to adjust concaves, lift them up a few times and drop so as to jar out dust. Wedging a block of wood between cylinder teeth and concaves will in some types of separator serve to bring up concaves when cylinder is slowly turned by hand.

There are from two to six rows of teeth in the concave, and usually the number of rows is adjustable or variable. Two rows will thresh oats, where six are required for flax and timothy. Four rows are commonly used for wheat and barley. The arrangement of rows of teeth and blanks is important. When four rows are used, one is commonly placed well back, one front, blank in the middle. When straw is dry and brittle, cylinder can be given "draw" by placing blank in front. Always use as few teeth and leave them as low as possible to thresh clean, since with more teeth than necessary set higher than required the straw will be cut up and a great deal of chopped straw will get into the sieves, all of which also requires additional power. Sometimes the teeth can be taken out of one row, so that one, three, or five rows may be used. For especially difficult grain like Turkey wheat, a concave with corrugated teeth may be used, in sets of three rows each up to nine rows. The corrugated teeth are used for alfalfa in localities where much is raised.

THE BEATER AND CHECK BOARD.

After the cylinder has loosened the grain from the husk and straw, it must still be separated. Some threshers have a grate under the cylinder and behind it. In any case the beater causes the heavy grain to work toward the bottom, and the check board keeps the grain from being carried to rear on top of the straw, where it would not have a chance to become separated. If the grain is very heavy or damp, there may be a tendency for the straw to stick to the cylinder and be carried around too far. In such a case the beater should be adjusted to give more space, and the check board raised to allow the straw to pass to the rear freely.

STRAW RACK.

The straw rack and conveyor carry the straw and grain to the rear with a vibratory movement, causing the grain to be shaken out. To do good work the straw rack must move with a sufficient number of vibrations per minute, say 230. A speed indicator on the crank shaft will show the number of vibrations best. Great care must be taken with this part of the thresher, or a great deal of grain will be carried into the straw. The less the straw is cut up, the better this portion of the machine works; so the smallest practicable number of teeth in the concave should be used.

The crank boxes and pitmans should be adjusted so that there is no pounding. If the rear vibrating arms drop too low they get below the dead center and are liable to break, at any rate causing severe pounding and hard running. To prevent this, the crank boxes can be moved forward by putting leather between them and the posts, or should be otherwise adjusted. The trouble being due to the pitmans having worn short, the pitmans may be lengthened in some way by putting pieces of leather over the end or the like, or new pitmans may be introduced.

THE FAN.

The chief difficulty likely to arise with the fan is blowing over grain. To prevent this blinds are usually arranged, which may be adjusted while the machine is running so as to prevent the grain from being blown over. At the same time it is important to clean the grain, so the adjustment should not go to one extreme or the other.

In windy weather the blinds should be closed more on one side than on the other. The speed of the fan must be adjusted to the requirements of the locality.

As much blast should be used as the grain will stand, and heavy feeding requires more wind than light feeding, since the chaff checks the blast to a certain extent.

Care should be taken that the wind board over the grain auger does not get bent, and it should be adjusted so that the strongest part of the blast will come about the middle of the sieve.

SIEVES.

There is usually one conveyor sieve, which causes the grain to move along, and shoe sieves, which are required to clean the grain thoroughly. Different kinds of sieves are provided for different kinds of grain, and the proper selection and adjustment of these sieves as to mesh, etc., is of the utmost importance.

Much depends on the way the sieves are set, and on the rate at which the thresher is fed, or the amount of work it is really doing. The best guide is close observation and experience, both your own and that of other threshermen.

CONVEYOR EXTENSION.

This carries the coarse chaff from the conveyor sieve to the stacker. The conveyor sieve should be coarse enough to let all the good grain through, as whatever is carried on to the extension must be returned with the tailings to the cylinder. This means so much waste work. The conveyor extension is removable, and should always be tight before machine is started. See that it is.

When necessary, the grain may be run over a screen, which differs from a sieve in that the mesh is small and intended to let dust and small chaff through while the grain does not pass. The refuse from the screen is dropped onto the ground. All screens have a tendency to become clogged, and in this condition obstruct the grain and wind. It is desirable not to use them except when necessary, and if used they should be frequently cleaned.

TAILINGS ELEVATOR.

The tailings are carried back to the cylinder by an elevator usually worked with a chain. This chain should be kept tight enough not to unhook, yet not so tight as to bind.

To put the chain into the elevator, tie a weight on a rope and drop it down the lower part of the elevator. The chain may be fastened to the rope and a man at the top can then pull the chain up, while another feeds it in at the bottom. When chain has been drawn up to the top, the rope should be dropped down upper portion of elevator and used at bottom to pull chain down after it has been adjusted over the sprocket. Some one at the bottom should continue to feed the chain in as it is pulled down, so that it will go into the elevator straight. When the chain has been pulled through it may be hooked and adjusted to lower sprocket, and tightened up by screws at top. Turn the chain around once by hand to make sure there are no kinks in it.

The tailings should be small, containing no light chaff and little full-size grain. They are a good indication of how the sieves are working. If much good grain is coming through, see if it gets over the conveyor sieve by way of the extension to the tailings auger, or over the shoe sieve. If the sieves are not right, they may be adjusted in various ways, according to the directions of the manufacturer.

Grain returned in the tailings is liable to get cracked in the cylinder, and much chaff in the tailings chokes the cylinder. For every reason, the tailings should be kept as low as possible.

SELF-FEEDER.

The self-feeder is arranged to cut the bands of the sheaves and feed the grain to the cylinder automatically.

It has a governor to prevent crowding in too much grain, and usually a change of pulleys for slow or fast feeding, as circumstances may require. In starting a new governor the friction pulley and inside of the band should have paint scraped off, and a little oil should be put on face of friction wheel. The carrier should not start till the machine attains full threshing motion, and to prevent this a few sheaves should be laid upon it. The knife arms should be raised or lowered to adjust them to the size of the sheaves and condition of the grain for cutting bands.

The cranks and carrier shaft boxes should be oiled regularly, but the friction bands should not be oiled after it once becomes smooth.

THE WIND STACKER.

The wind stacker is arranged to swing by a hand-wheel or the like, and also automatically.

Great care should be taken not to use the hand moving apparatus when the stacker is set for automatic moving, as a break is liable to follow. There is a clutch to stop the stacker, however. At times it will be more convenient to leave off the belt that causes the automatic movement.

By the use of various pulleys the speed of the stacker may be altered, and it should be run no faster than is necessary to do the work required, which will depend on the character of the straw. Any extra speed used will add to the cost of running the engine and is a loss in economy.

In moving machine with wind stacker in place, care should be taken to see that it rests in its support before machine moves.

The canvas curtain under the decking, used to turn the straw into the hopper, may need a piece of wood fastened to its lower edge to keep it more stiff when stiff rve straw is passing. The bearings of the fan and jack shafts should be kept well lubricated with hard oil, and the bevel gears should be kept well greased with axle grease applied with a stick. Other bearings and worm gear of automatic device should be oiled with soft oil.

The attached stacker is simple in operation, and if it is desired not to use the automatic swinging device but swing by hand, the automatic gear may be thrown out. An independent stacker is managed in much the same way.

ATTACHMENTS.

A weigher, bagger, and a high loader are usually used with a separator. Their operation is simple, and depends upon the particular type or make.

BELTING.

The care of the belting is one of the most important things about the management of a threshing machine, and success or failure will depend largely on the condition in which the belts are kept. Of course the hair side should be run next the band wheel. Once there was disagreement among engineers on this point, but it has been conclusively proven that belts wear longer this way and get better friction, for the simple reason that the flesh side is more flexible than the hair side, and when on the outside better accommodates itself to the shape of the pulley. If the hair side is outermost, it will be stretched more or less in going around the pulley and in time will crack. Rubber belts must be run with the seam on the outside.

When leather belts become hard they should be softened with neatsfoot oil. A flexible belt is said to transmit considerably more power than a hard one.

Pulleys must be kept in line or the belt will slip off. When pulleys are in line the belt has a tendency to work to the tightest point. Hence pulleys are usually made larger in the middle, which is called "crowning."

Belts on a separator should be looked over every day, and when any lacing is worn, it should be renewed at once. This will prevent breaks during working, with loss of time. Some threshermen carry an extra set of belts to be ready in case anything does break, and they assert that they save money by so doing.

Lacing is not stronger in proportion as it is heavy. If it is heavy and clumsy it gets strained in going round

the pulley, and soon gives out. The ideal way to lace
a belt is to make it as nearly like the rest of the belt as
possible, so that it will go over the pulleys without a
jar. The ends of the belt should be cut off square with
a try square, and a small punch used for making holes.
Holes should be equally spaced, and outside ones not so
near the edge as to tear out. The rule is a hole to every
inch of the belt, and in a leather belt they may be as
close as a quarter of an inch to the ends without tearing
out. Other things being equal, the nearer the ends the
holes are the better, as belt will then pass over pulley
more easily. The chief danger of tearing is between
the holes.

A stacker web belt may be laced by turning the ends
up and lacing them together flat at right angles to rest
of belt. Rubber or cotton belting that does not run over
idler or tightener pulleys so that both sides must be
smooth may be laced in this way. This lacing lasts two
or three times as long with such belts as any other, for
the reason that the string is not exposed to wear and
there is no straining in passing round pulleys.

The ordinary method of lacing a leather belt is to
make the laces straight on the pulley side, all running
in the same direction as the movement of the belt, and
crossing them on the outside diagonally in both directions. When belts run on pulleys on both sides, as they
do on the belt driving beater and crank, and also on
wind stacker, a hinge lacing may be made by crossing
the lacing around the end of the belt to the next adjacent
hole opposite, the lacing showing the same on both
sides. This allows the belt to bend equally well either
way.

The best way to fasten a lacing is to punch a hole
where the next row of lace holes would come when the
belt is cut off, and after passing the lace through this
hole, bring the end around and force it through again,
cutting the end off short after it has passed through.
This hole must be small enough to hold the lace securely,
and care should be taken that it is in position to be used
as a lace-hole the next time a series of holes is required.

New belts stretch a good deal, and the ends of the

lacing should not be cut off short till the stretch is taken out of the belts.

Belting that has got wet will shrink and lacing must be let out before belt is put on again. Tight belts have been known to break the end of a shaft off, and always cause unnecessary friction.

Cotton or Gandy belting should not be punched for lacing, but holes made with a pointed awl, since punching cuts some of the threads and weakens belt.

HOW TO BECOME A GOOD FEEDER.

The art of becoming a good feeder will not be learned in a day. The bundles should be tipped well up against the cylinder cap, and flat bundles turned on edge, so that cylinder will take them from the top. It is not hard to spread a bundle, and in fast threshing a bundle may be fed on each side, each bundle being kept pretty well to its own side, while the cylinder is kept full the entire width. A good feeder will keep the straw carrier evenly covered with straw, and will watch the stacker, tailings and grain elevator and know the moment anything goes wrong.

WASTE.

No threshing machine will save every kernel of the grain, but the best results can be attained only by care and judgment in operating.

It is easy to exaggerate the loss of grain, for if a very small stream of grain is seen going into the straw it will seem enormous, though it will not amount to a bushel a day. There are practically a million kernels of wheat in a bushel, or 600 handfuls, and even if a handful is wasted every minute, it would not be enough to counterbalance the saving in finishing a job quickly.

Of course, waste must be watched, however, and checked if too great. First determine whether the grain is carried over in the straw or the waste is at the shoe sieve.

If the waste is in the conveyor sieve, catch a handful of the chaff, and if grain is found, see whether the sieve is the proper mesh. Too high a speed will cause the grain to be carried over. If too many teeth are used in the concave, the conveyor sieve will be forced to carry more chaff than it can handle. The blast may be too strong and carry over grain, so adjust the blinds that the blast will be no stronger than is necessary to clean the wheat well and keep sieves free. If grain is still carried over, the conveyor sieve may be adjusted for more open work, but care should be taken not to overwork the shoe sieve. Be careful that the wind board is not bent so that some grain will go into the fan and be thrown out of the machine altogether.

If the grain is not separated from the straw thoroughly, it may be due to "slugging" the cylinder (result of poor feeding), causing a variable motion. It may also be because speed of crank is not high enough. Check board should be adjusted as low as possible to prevent grain being carried on top of straw. See that cylinder and concave teeth are properly adjusted so as not to cut up straw, while at the same time threshing out all the grain. Sometimes heads not threshed out by the cylinder will be threshed out by the fan of the wind stacker, and the fault will be placed on the separating portions instead of on the imperfect cylinder.

Grain passes through the cylinder at the rate of about a mile a minute. The beater reduces this to 1,500 feet per minute. After passing the check board the straw moves about 36 feet per minute. At these three different speeds the straw passes the 17 feet length of the machine in about 25 seconds. The problem is to stop the grain while the straw is allowed to pass out. Evidently there must be a small percentage of loss, and there is always a limit as to what it will pay to try to save. Each man must judge for himself.

BALANCING A CYLINDER.

A cylinder should be so balanced that it will come to rest at any point. In a rough way a cylinder may be balanced by placing the journals on two carpenter's squares laid on saw-horses. Gently roll the cylinder back and forth and every time it stops, make a chalk mark on the uppermost bar. If the same bar comes up three times in succession it probably is light, and a wedge should be driven under center band at chalk mark. Continue experimenting until cylinder will come to rest at any point.

COVERING PULLEYS.

This is easily done, but care must be taken that the leathers are tight or they will soon come off.

To cover a cylinder pulley, take off what remains of the old cover, pull out the nails, and renew the wedges if necessary. Select a good piece of leather a little wider than face of pulley and about four inches longer than enough to go around. Soak it in water for about an hour. Cut one end square and nail it to the wedges, using nails just long enough to clinch. Put a clamp made of two pieces of wood and two bolts on the leather, block the cylinder to keep it from turning, and by means of two short levers pry over the clamp to stretch the leather. Nail to the next wedges, move the clamp and nail to each in turn, finally nailing to the first one again before cutting off. Trim the edges even with the rim of the pulley.

The same method may be used with riveted covers.

CARE OF A SEPARATOR.

A good separator ought to last ten years, and many have been in use twice that time. After the season is over the machine ought to be thoroughly cleaned and stored in a dry place. Dirt on a machine holds moisture and will ruin a separator during a winter if it is left on. It also causes the wood to rot and sieves and iron work to rust.

Once in two years at least a separator ought to have a good coat of first-class coach varnish. Before varnishing, clean off all grease and oil with benzine and see that paint is bright.

At the beginning of the season give the machine a thorough overhauling, putting new teeth in cylinder if any are imperfect, and new slats in stacker web or straw rack if they are needed. Worn boxes should be taken up or rebabbitted, and conveyor and shoe eccentrics replaced if worn out. Tighten nuts, replace lost bolts, leaving the nut always turned square with the piece it rests on. Every separator ought to be covered with a canvas during the season. It will pay.

The right and left sides of a threshing machine are reckoned from the position of the feeder as he stands facing the machine.

In case of fire, the quickest way is to let the engine pull the machine out by the belt. Take blocks away from wheels, place a man at end of tongue to steer, and back engine slowly. If necessary, men should help the wheels to start out of holes or soft places.

Watch the forks of the pitchers to see that none are loose on the handles, especially if a self-feeder is used. A pitchfork in a separator is a bad thing.

CHAPTER XV.

Q. If you were called on to take charge of a plant what would be your first duty?

A. To ascertain the exact condition of the boiler and all its attachments (safety valve, steam gauge, pump, injector), and engine.

Q. How often would you blow off and clean your boilers if you had ordinary water to use?

A. Once a month.

Q. What steam pressure will be allowed on a boiler 50 inches diameter ⅜ inch thick, 60,000 T. S. 1-6 of tensile strength factor of safety?

A. One-sixth of tensile strength of plate multiplied by thickness of plate, divided by one-half of the diameter of boiler, gives safe working pressure.

Q. How much heating surface is allowed per horse power by builders of boilers?

A. Twelve to fifteen feet for tubular and flue boilers.

Q. How do you estimate the strength of a boiler?

A. By its diameter and thickness of metal.

Q. Which is the better, single or double riveting?

A. Double riveting is from sixteen to twenty per cent stronger than single.

Q. How much grate surface do boiler makers allow per horse power?

A. About two-thirds of a square foot.

Q. Of what use is a mud drum on a boiler, if any?

A. For collecting all the sediment of the boiler.

Q. How often should it be blown out?

A. Three or four times a day.

*Furnished by courtesy of a friend of Aultman & Taylor Co.

Q. Of what use is a steam dome on a boiler?
A. For storage of dry steam.
Q. What is the object of a safety valve on a boiler?
A. To relieve pressure.
Q. What is your duty with reference to it?
A. To raise it twice a day and see that it is in good order.
Q. What is the use of check valve on a boiler?
A. To prevent water from returning back into pump or injector which feeds the boiler.
Q. Do you think a man-hole in the shell on top of a boiler weakens it any?
A. Yes, to a certain extent.
Q. What effect has cold water on hot boiler plates?
A. It will fracture them.
Q. Where should the gauge cock be located?
A. The lowest gauge cock ought to be placed about an inch and a half above the top row of flues.
Q. How would you have your blow-off located?
A. In the bottom of mud-drum or boiler.
Q. How would you have your check valve arranged?
A. With a stop cock between check and boiler.
Q. How many valves are there in a common plunger force pump?
A. Two or more—a receiving and a discharge valve.
Q. How are they located?
A. One on the suction side, the other on the discharge.
Q. How do you find the proper size of safety valves for boilers?
A. Three square feet of grate surface is allowed for one inch area of spring loaded valves; or two square feet of grate surface to one inch area of common lever valves.
Q. Give the reasons why pumps do not work sometimes?
A. Leak in suction, leak around the plunger, leaky check valve, or valves out of order, or lift too long.
Q. How often ought boilers to be thoroughly examined and tested?

A. Twice a year.

Q. How would you test them?

A. With hammer and with hydrostatic test, using warm water.

Q. Describe the single acting plunger pump; how it gets and discharges its water?

A. The plunger displaces the air in the water pipe, causing a vacuum which is filled by the atmosphere forcing the water therein; the receiving valve closes and the plunger forces the water out through the discharge valve.

Q. What is the most economical boiler-feeder?

A. The (Trix) Exhaust Injector.*

Q. What economy is there in the Exhaust Injector?

A. From 15 to 25 per cent saving in fuel.

Q. Where is the best place to enter the boiler with the feed water?

A. Below the water level, but so that the cold water can not strike hot plates. If injector is used this is not so material as feed water is always hot.

Q. What are the principal causes of priming in boilers?

A. To high water, not steam room enough, misconstruction, engine too large for boiler.

Q. How do you keep boilers clean or remove scale therefrom?

A. The best "scale solvent" and "feed water purifier" is an honest, intelligent engineer who will regularly open up his boilers and clean them thoroughly, soaking boilers in rain water now and then.

Q. If you found a thin plate, what would you do?

A. Put a patch on it.

Q. Would you put it on the inside or outside?

A. Inside.

Q. Why so?

A. Because the action that has weakened the plate will then set on the patch, and when this is worn it can be repeated.

---

*So says one expert. Others may think otherwise.

Q. If you found several thin places, what would you do?

A. Patch each and reduce the pressure.

Q. If you found a blistered plate?

A. Put a patch on the fire side.

Q. If you found a plate on the bottom buckled?

A. Put a stay through the center of buckle.

Q. If you found several of the plates buckled?

A. Stay each and reduce the pressure.

Q. What is to be done with a cracked plate?

A. Drill a hole at each end of crack, caulk the crack and put a patch over it.

Q. How do you change the water in the boiler when the steam is up?

A. By putting on more feed and opening the surface blow cock.

Q. If the safety valve was stuck, how would you relieve the pressure on the boiler if the steam was up and could not make its escape?

A. Work the steam off with engine after covering fires heavy with coal or ashes, and when the boiler is sufficiently cool put safety valve in working order.

Q. If water in boiler is suffered to get too low, what may be the result?

A. Burn top of combustion chamber and tubes, perhaps cause an explosion.

Q. If water is allowed to get too high, what result?

A. Cause priming, perhaps cause breaking of cylinder covers or heads.

Q. What are the principal causes of foaming in boilers?

A. Dirty and impure water.

Q. How can foaming be stopped?

A. Close throttle and keep closed long enough to show true level of water. If that level is sufficiently high, feeding and blowing off will usually suffice to correct the evil.

Q. What would you do if you should find your water gone from sight very suddenly?

A. Draw the fires and cool off as quickly as possible.

Never open or close any outlets of steam when your water is out of sight.

Q. What precautions should you take to blow down a part of the water in your boiler while running with a good fire?

A. Never leave the blow-off valve, and watch the water level.

Q. How much water would you blow off at once while running?

A. Never blow off more than one gauge of water at a time while running.

Q. What general views have you in regard to boiler explosions—what is the greatest cause?

A. Ignorance and neglect are the greatest causes of boiler explosions.

Q. What precaution should the engineer take when necessary to stop with heavy fires?

A. Close dampers, put on injector or pump and if a bleeder is attached, use it.

Q. Where is the proper water level in boilers?

A. A safe water level is about two and a half inches over top row of flues.

Q. What is an engineer's first duty on entering the boiler room?

A. To ascertain the true water level.

Q. When should a boiler be blown out?

A. After it is cooled off, never while hot.

Q. When laying up a boiler what should be done?

A. Clean thoroughly inside and out; remove all oxidation and paint places with red lead; examine all stays and braces to see if any are loose or badly worn.

Q. What is the last thing to do at night before leaving plant?

A. Look around for greasy waste, hot coals, matches, or anything which could fire the building.

Q. What would you do if you had a plant in good working order?

A. Keep it so, and let well enough alone.

Q. Of what use is the indicator?

A. The indicator is used to determine the indicated

power developed by an engine, to serve as a guide in setting valves and showing the action of the steam in the cylinder.

Q. How would you increase the power of an engine?

A. To increase the power of an engine, increase the speed; or get higher pressure of steam, use less expansion.

Q. How do you find the horsepower of an engine?

A. Multiply the speed of piston in feet per minute by the total effective pressure upon the piston in pounds and divide the product by 33,000.

Q. Which has the most friction, a perfectly fitted, or an imperfectly fitted valve or bearing?

A. An imperfect one.

Q. How hot can you get water under atmospheric pressure with exhaust steam?

A. 12 degrees.

Q. Does pressure have any influence on the boiling point?

A. Yes.

Q. Which do you think is the best economy, to run with your throttle wide open or partly shut?

A. Always have the throttle wide open on a governor engine.

Q. At what temperature has iron the greatest tensile strength?

A. About 600 degrees.

Q. In what position on the shaft does the eccentric stand in relation to the crank?

A. The throw of the eccentric should always be in advance of the crank pin.

Q. About how many pounds of water are required to yield one horsepower with our best engines?

A. From 25 to 30.

Q. What is meant by atmospheric pressure?

A. The weight of the atmosphere.

Q. What is the weight of atmosphere at sea level?

A. 14.7 pounds.

Q. What is the coal consumption per hour per indicated horsepower?

A.   Varies from one and a half to seven pounds.

Q.   What is the consumption of coal per hour on a square foot of grate surface?

A.   From 10 to 12 pounds.

Q.   What is the water consumption in pounds per hour per indicated horsepower?

A.   From 25 to 60 pounds.

Q.   How many pounds of water can be evaporated with one pound of best soft coal?

A.   From 7 to 10 pounds.

Q.   How much steam will one cubic inch of water evaporate under atmospheric pressure?

A.   One cubic foot of steam (approximately).

Q.   What is the weight of a cubic foot of fresh water?

A.   Sixty-two and a half pounds.

Q.   What is the weight of a cubic foot of iron?

A.   486.6 pounds.

Q.   What is the weight of a square foot of one-half inch boiler plate?

A.   20 pounds.

Q.   How much wood equals one ton of soft coal for steam purposes?

A.   About 4,000 pounds of wood.

Q.   How long have you run engines?

Q.   Have you ever done your own firing?

Q.   What is the source of all power in the steam engine?

A.   The heat stored up in the coal.

Q.   How is the heat liberated from the coal?

A.   By burning it; that is, by combustion.

Q.   Of what does coal consist?

A.   Carbon, hydrogen, nitrogen, sulphur, oxygen and ash.

Q.   What are the relative proportions of these that enter into coal?

A.   There are different proportions in different specimens of coal, but the following shows the average per cent: Carbon, 80; hydrogen, 5; nitrogen, 1; sulphur, 2; oxygen, 7; ash, 5.

Q. What must be mixed with coal before it will burn?

A. Atmospheric air.

Q. What is air composed of?

A. It is composed of nitrogen and oxygen in the proportion of 77 of nitrogen to 23 of oxygen.

Q. What parts of the air mix with what parts of the coal?

A. The oxygen of the air mixes with the carbon and hydrogen of the coal.

Q. How much air must mix with the coal?

A. 150 cubic feet of air for every pound of coal.

Q. How many pounds of air are required to burn one pound of carbon?

A. Twelve.

Q. How many pounds of air are required to burn one pound of hydrogen?

A. Thirty-six.

Q. Is hydrogen hotter than carbon?

A. Yes, four and one-half times hotter.

Q. What part of the coal gives out the most heat?

A. The hydrogen does part for part, but as there is so much more of carbon than hydrogen in the coal we get the greatest amount of heat from carbon.

Q. In how many different ways is heat transmitted?

A. Three; by radiation, by conduction and by convection.

Q. If the fire consisted of glowing fuel, show how the heat enters the water and forms steam?

A. The heat from the glowing fuel passes by radiation through the air space above the fuel to the furnace crown. There it passes through the iron of the crown by conduction. There it warms the water resting on the crown, which then rises and parts with its heat to the colder water by conduction till the whole mass of water is heated. Then the heated water rises to the surface and parts with its steam, so a constant circulation of water is maintained by convection.

Q. What does water consist of?

A. Oxygen and hydrogen.

Q.   In what proportion?

A.   Eight of oxygen to one of hydrogen by weight.

Q.   What are the different kinds of heat?

A.   Latent heat, sensible heat and sometimes total heat.

Q.   What is meant by latent heat?

A.   Heat that does not affect the thermometer and which expands itself in changing the nature of a body, such as turning ice into water or water into steam.

Q.   Under what circumstances do bodies get latent heat?

A.   When they are passing from a solid state to a liquid or from a liquid to a gaseous state.

Q.   How can latent heat be recovered?

A.   By bringing the body back from a state of gas to a liquid or from that of a liquid to that of a solid.

Q.   What is meant by a thermal unit?

A.   The heat necessary to raise one pound of water at 39 degrees Fn. 1 degree Fahrenheit.

Q.   If the power is in coal, why should we use steam?

A.   Because steam has some properties which make it an invaluable agent for applying the energy of the heat to the engine.

Q.   What is steam?

A.   It is an invisible elastic gas generated from water by the application of heat.

Q.   What are its properties which make it so valuable to us?

A.   1.—The ease with which we can condense it. 2.—Its great expansive power. 3.—The small space it occupies when condensed.

Q.   Why do you condense the steam?

A.   To form a vacuum and so destroy the back pressure that would otherwise be on the piston and thus get more useful work out of the steam.

Q.   What is vacuum?

A.   A space void of all pressure.

Q.   How do you maintain a vacuum?

A.   By the steam used being constantly condensed by the cold water or cold tubes, and the air pump as constantly clearing the condenser out.

Q. Why does condensing the used steam form a vacuum?

A. Because a cubic foot of steam, at atmospheric pressure, shrinks into about a cubic inch of water.

Q. What do you understand by the term horse power?

A. A horse power is equivalent to raising 33,000 pounds one foot per minute, or 550 pounds raised one foot per second.

Q. How do you calculate the horse power of tubular or flue boilers?

A. For tubular boilers, multiply the square of the diameter by length, and divide by four. For flue boilers, multiply the diameter by the length and divide by four; or, multiply area of grate surface in square feet by $1\frac{1}{2}$.

Q. What do you understand by lead on an engine's valve?

A. Lead on a valve is the admission of steam into the cylinder before the piston completes its stroke.

Q. What is the clearance of an engine as the term is applied at the present time?

A. Clearance is the space between the cylinder head and the piston head with the ports included.

Q. What are considered the greatest improvements on the stationary engine in the last forty years?

A. The governor, the Corliss valve gear and the triple compound expansion.

Q. What is meant by triple expansion engine?

A. A triple expansion engine has three cylinders using the steam expansively in each one.

Q. What is a condenser as applied to an engine?

A. The condenser is a part of the low pressure engine and is a receptacle into which the exhaust enters and is there condensed.

Q. What are the principles which distinguish a high pressure from a low pressure engine?

A. Where no condenser is used and the exhaust steam is open to the atmosphere.

Q. About how much gain is there by using the condenser?

A.   17 to 25 per cent where cost of water is not figured.

Q.   What do you understand by the use of steam expansively?

A.   Where steam admitted at a certain pressure is
cut off and allowed to expand to a lower pressure.

Q.   How many inches of vacuum give the best results in a condensing engine?

A.   Usually considered 25.

Q.   What is meant by a horizontal tandem engine?

A.   One cylinder being behind the other with two
pistons on same rod.

Q.   What is a Corliss valve gear?

A.   (*Describe the half moon or crab claw gear, or
oval arm gear with dash pots.*)

Q.   From what cause do belts have the power to
drive shafting?

A.   By friction or cohesion.

Q.   What do you understand by lap?

A.   Outside lap is that portion of valve which extends beyond the ports when valve is placed on the
center of travel, and inside lap is that portion of valve
which projects over the ports on the inside or towards
the middle of valve.

Q.   What is the use of lap?

A.   To give the engine compression.

Q.   Where is the dead center of an engine?

A.   The point where the crank and the piston rod
are in the same right line.

Q.   What is the tensile strength of American boiler
iron?

A.   40,000 to 60,000 pounds per square inch.

Q.   What is very high tensile strength in boiler iron
apt to go with?

A.   Lack of homogeneousness and lack of toughness.

Q.   What is the advantage of toughness in boiler
plate?

A.   It stands irregular strains and sudden shocks better.

Q.   What are the principal defects found in boiler
iron?

A.   Imperfect welding, brittleness, low ductility.

Q.   What are the advantages of steel as a material for boiler plates?

A.   Homogeneity, tensile strength, malleability, ductility and freedom from laminations and blisters.

Q.   What are the disadvantages of steel as a material for boiler plates?

A.   It requires greater skill in working than iron, and has, as bad qualities, brittleness, low ductility and flaws induced by the pressure of gas bubbles in the ingot.

Q.   When would you oil an engine?

A.   Before starting it and as often while running as necessary.

Q.   How do you find proper size of any stay bolts for a well made boiler?

A.   First, multiply the given steam pressure per square inch by the square of the distance between centers of stay bolts, and divide the product by 6,000, and call the answer "the quotient." Second, divide "the quotient" by .7854, and extract the square root of the last quotient; the answer will give the required diameter of stay bolts at the bottom of thread.

Q.   In what position would you place an engine, to take up any slack motion of the reciprocating parts?

A.   Place engine in the position where the least wear takes place on the journals. That is, in taking up the wear of the crank-pin brasses, place the engine on either dead center, as, when running, there is but little wear upon the crank-pin at these points. If taking up the cross-head pin brasses—without disconnecting and swinging the rod—place the engine at half stroke, which is the extreme point of swing of the rod, there being the least wear on the brasses and cross-head pin in this position.

Q.   What benefits are derived by using flywheels on steam engines?

A.   The energy developed in the cylinder while the steam is doing its work is stored up in the flywheel, and given out by it while there is no work being done in the cylinder—that is, when the engine is passing the dead centers. This tends to keep the speed of the engine shaft steady.

Q. Name several kinds of reducing motions, as used in indicator practice?

A. The pantograph, the pendulum, the brumbo pulley, the reducing wheel.

Q. How can an engineer tell from an indicator diagram whether the piston or valves are leaking?

A. Leaky steam valves will cause the expansion curve to become convex; that is, it will not follow hyperbolic expansion, and will also show increased back pressure. But if the exhaust valves leak also, one may offset the other, and the indicator diagram would show no leak.

A leaky piston can be detected by a rapid falling in the pressure on the expansion curve immediately after the point of cut-off. It will also show increased back pressure.

A falling in pressure in the upper portion of the compression curve shows a leak in the exhaust valve.

Q. What would be the best method of treating a badly scaled boiler, that was to be cleaned by a liberal use of compound?

A. First open the boiler up and note where the loose scale, if any, has lodged. Wash out thoroughly and put in the required amount of compound. While the boiler is in service, open the blow-off valve for a few seconds, two or three times a day, to be assured that it does not become stopped up with scale.

After running the boiler for a week, shut it down, and, when the pressure is down and the boiler cooled off, run the water out and take off the hand-hole plates. Note what effect the compound has had on the scale, and where the disengaged scale has lodged. Wash out thoroughly and use judgment as to whether it is advisable to use a less or greater quantity of compound, or to add a small quantity daily.

Continue the washing out at short intervals, as many boilers have been burned by large quantities of scale dropping on the crown sheets and not being removed.

Q. If a condenser was attached to a side-valve engine, that had been set to run non-condensing, what changes, if any, would be necessary?

A.   More lap would have to be added to the valve to cut off the steam at an earlier point of the stroke; if not, the initial pressure into the cylinder would be throttled down and the economy, to be gained from running condensing, lessened.

Q.   If you are carrying a vacuum equal to 27½ inches of mercury, what should the temperature of the water in the hot well be?

A.   108 degrees Fahrenheit.

Q.   Define specific gravity.

A.   The specific gravity of a substance is the number which expresses the relation between the weights of equal volume of that substance, and distilled water of 60 degrees Fahrenheit.

Q.   Find the specific gravity of a body whose volume is 12 cubic inches, and which floats in water with 7 cubic inches immersed.

A.   When a body floats in water, it displaces a quantity of water equal to the weight of the floating body. Thus, if a body of 12 cubic inches in volume floats with 7 cubic inches immersed, 7 cubic inches of water must be equal in weight to 12 cubic inches of the substance and one cubic inch of water to twelve-sevenths cubic inches of the substance.

As specific gravity equals weight of one volume of substance divided by weight of equal volume of water, then specific gravity of the substance in this case equals 1 divided by twelve-sevenths.

USEFUL INFORMATION.

To find circumference of a circle, multiply diameter by 3.1416.

To find diameter of a circle, multiply circumference by .31831.

To find area of a circle multiply square of diameter by .7854.

To find area of a triangle, multiply base by one-half the perpendicular height.

To find surface of a ball, multiply square of diameter by 3.1416.

To find solidity of a sphere, multiply cube of diameter by .5236.

To find side of an equal square, multiply diameter by 8862.

To find cubic inches in a ball multiply cube of diameter by .5236.

Doubling the diameter of a pipe increases its capacity four times.

A gallon of water (U. S. standard) weighs 8 1-3 pounds and contains 231 cubic inches.

A cubic foot of water contains 7½ gallons, 1728 cubic inches, and weighs 62½ pounds.

To find the pressure in pounds per square inch of a column of water multiply the height of the column in feet by .434.

Steam rising from water at its boiling point (212 degrees) has a pressure equal to the atmosphere (14.7 pounds to the square inch).

A standard horse power: The evaporation of 30 lbs. of water per hour from a feed water temperature of 100 degrees F. into steam at 70 lbs. gauge pressure.

To find capacity of tanks any size; given dimensions of a cylinder in inches, to find its capacity in U. S. gallons: Square the diameter, multiply by the length and by .0034.

To ascertain heating surface in tubular boilers, multiply two-thirds of the circumference of boiler by length of boiler in inches and add to it the area of all the tubes.

One-sixth of tensile strength of plate multiplied by thickness of plate and divided by one-half the diameter of boiler gives safe working pressure for tubular boilers. For marine boilers add 20 per cent for drilled holes.

To find the horsepower of an engine, the following four factors must be considered: Mean effective or average pressure on the cylinder, length of stroke, diameter of cylinder, and number of revolutions per minute. Find the area of the piston in square inches by multiplying the diameter by 3.1416 and multiply the result by the steam pressure in pounds per square inch; multiply this product by twice the product of the length of the stroke in feet and the number of revolutions per minute; divide the result by 33,000, and the result will be the horsepower of the engine.

(Theoretically a horsepower is a power that will raise 33,000 pounds one foot in one minute.)

The power of fuel is measured theoretically from the following basis: If a pound weight fall 780 feet in a vacuum, it will generate heat enough to raise the temperature of one pound of water one degree. Conversely, power that will raise one pound of water one degree in temperature will raise a one pound weight 780 feet. The heat force required to turn a pound of water at 32 degrees into steam would lift a ton weight 400 feet high, or develop two-fifths of one horsepower for an hour. The best farm engine practically uses 35 pounds of water per horsepower per hour, showing that one pound of water would develop only one-thirty-fifth of a horsepower in an hour, or 7 1-7 per cent of the heat force liberated. The rest of the heat force is lost in various ways, as explained in the body of this book.

The following* will assist in determining the amount of power supplied to an engine:

"For instance, a 1 inch belt of the standard grade with the proper tension, neither too tight or too loose, running at a maximum speed of 800 feet a minute will transmit one horsepower, running 1,600 feet two horsepower and 2,400 feet three horsepower. A 2-inch belt at the same speed, twice the power.

"Now if you know the circumference of your flywheel, the number of revolutions your engine is making and the width of belt, you can figure very nearly the amount of power you can supply without slipping your belt. For instance, we will say your flywheel is 40 inches in diameter or 10.5 feet nearly in circumference and your engine was running 225 revolutions a minute, your belt would be traveling 225x10.5 feet = 2362.5 feet, or very nearly 2,400 feet, and if one inch of belt would transmit three

*J. H. Maggard in "Rough and Tumble Engineering."

horsepower running this speed, a 6-inch belt would transmit eighteen horsepower, a 7-inch belt twenty-one horsepower, an 8-inch belt twenty-four horsepower, and so on. With the above as a basis for figuring you can satisfy yourself as to the power you are furnishing. To get the best results a belt wants to sag slightly, as it hugs the pulley closer, and will last much longer."

KEYING PULLEYS.*

A key must be of equal width its whole length and accurately fit the seats on shaft and in pulley. The thickness should vary enough to make the taper correspond with that of the seat in the pulley. The keys should be driven in tight enough to be safe against working loose. The hubs of most of the pulleys on the machine run against the boxes, and in keying these on, about 1-32 of an inch end play to the shaft should be allowed, because there is danger of the pulley rubbing so hard against the end of the box as to cause it to heat.

A key that is too thin but otherwise fits all right can be made tight by putting a strip of tin between the key and the bottom of the seat in the pulley.

Drawing Keys. If a part of the key stands outside of the hub, catch it with a pair of horseshoe pinchers and pry with them against the hub, at the same time hitting the hub with a hammer so as to drive pulley on. A key can sometimes be drawn by catching the end of it with a claw hammer and driving on the hub of pulley. If pulley is against box and key cut off flush with hub, take the shaft out and use a drift from the inside, or if seat is not long enough to make this possible, drive the pulley on until the key loosens.

BABBITTING BOXES.*

To babbitt any kind of a box, first chip out all of the old babbitt and clean the shaft and box thoroughly with benzine. This is necessary or gas will be formed from the grease when the hot metal is poured in and leave "blow holes." In babbitting a *solid box* cover the shaft

*Courtesy J. I. Case Threshing Machine Co., from "Science of Successful Threshing."

with paper, draw it smooth and tight, and fasten the lapped ends with mucilage. If this is not done the shrinkage of the metal in cooling will make it fast on the shaft, so that it can't be moved. If this happened it would be necessary to put the shaft and box together in the fire and melt the babbitt out or else break the box to get it off. Paper around the shaft will prevent this and if taken out when the babbitt has cooled the shaft will be found to be just tight enough to run well.

Before pouring the box, block up the shaft until it is in line and in center of the box and put stiff putty around the shaft and against the ends of the box to keep the babbitt from running out. Be sure to leave air-holes at each end at the top, making a little funnel of putty around each. Also make a larger funnel around the pouring hole, or, if there is none, enlarge one of the air-holes at the end and pour in that. The metal should be heated until it is just hot enough to run freely and the fire should not be too far away. When ready to pour the box, don't hesitate or stop, but pour continuously and rapidly until the metal appears at the air holes. The oil hole may be stopped with a wooden plug and if this plug extends through far enough to touch the shaft, it will leave a hole through the babbitt so that it will not be necessary to drill one.

A split box is babbitted in the same manner except that strips of cardboard or sheet-iron are placed between the two halves of the box and against the shaft to divide the babbitt. To let the babbitt run from the upper half to the lower, cut four or six V-shaped notches, a quarter of an inch deep, in the edges of the sheet-iron or cardboard that come against the shaft. Cover the shaft with paper and put cardboard liners between the box to allow for adjustment as it wears. Bolt the cap on securely before pouring. When the babbitt has cooled, break the box apart by driving a cold chisel between the two halves. Trim off the sharp edges of the babbitt and with a round-nose chisel cut oil grooves from the oil hole towards the ends of the box and on the slack side of the box or the one opposite to the direction in which the belt pulls.

The ladle should hold six or eight pounds of metal. If much larger it is awkward to handle and if too small it will not keep the metal hot long enough to pour a good box. The cylinder boxes on the separator take from two to three pounds of metal each. If no putty is at hand, clay mixed to the proper consistency may be used. Use the best babbitt you can get for the cylinder boxes. If not sure of the quality, use ordinary zinc. It is not expensive and is generally satisfactory.

MISCELLANEOUS.

Lime may be taken out of an injector by soaking it over night in a mixture of one part of muriatic acid and ten parts soft water. If a larger proportion of acid is used it is likely to spoil the injector.

A good blacking for boilers and smokestacks is asphaltum dissolved in turpentine.

To polish brass, dissolve 5 cents' worth of oxalic acid in a pint of water and use to clean the brass.. When tarnish has been removed, dry and polish with chalk or whiting.

It is said that iron or steel will not rust if it is placed for a few minutes in a warm solution of washing soda.

Grease on the bottom of a boiler will stick there and prevent the water from conducting away the heat. When steel is thus covered with grease it will soon melt in a hot fire, causing a boiler to burst if the steel is poor, or warping it out of shape if the steel is good.

Sulphate of lime in water, causing scale, may be counteracted and scale removed by using coal oil and sal soda. When water contains carbonate of lime, molasses will remove the scale.

CODE OF WHISTLE SIGNALS.

One short sound means to stop.

Two short sounds means the engine is about to begin work.

Three medium short sounds mean that the machine will soon need grain and grain haulers should hurry.

One rather long sound followed by three short ones means the water is low and water hauler should hurry.

A succession of short, quick whistles means distress or fire.

WEIGHT PER BUSHEL OF GRAIN.

The following table gives the number of pounds per bushel required by law or custom in the sale of grain in the several states:

	Barley	Beans	Buckwheat	Clover	Flax	Millet	Oats	Rye	Shelled Corn	Timothy	Wheat
Arkansas	48	60	52	60	..	..	..	56	56	45	60
California	50	..	40	..	..	..	32	54	52	..	60
Connecticut	..	..	45	..	..	..	32	56	56	..	56
District of Columbia	47	62	48	60	..	..	32	56	56	45	60
Georgia	40	..	..	60	..	..	35	56	56	45	60
Illinois	48	60	52	60	56	45	32	56	56	..	60
Indiana	48	60	50	60	..	..	32	56	56	45	60
Iowa	48	60	52	60	56	48	32	56	56	45	60
Kansas	50	60	50	..	..	..	32	56	56	45	60
Kentucky	48	60	52	60	56	..	32	56	56	45	60
Louisiana	32	..	..	..	..	..	32	..	56	..	60
Maine	48	64	48	..	..	..	30	..	56	..	60
Manitoba	48	..	48	60	56	34	..	56	56	..	60
Maryland	48	64	48	..	..	..	32	56	56	45	60
Massachusetts	48	48	..	..	..	..	32	56	56	..	60
Michigan	48	..	48	60	56	..	32	56	56	45	60
Minnesota	48	60	42	60	..	48	32	56	56	..	60
Missouri	48	60	52	60	56	50	32	56	56	45	60
Nebraska	48	60	52	60	..	..	34	56	56	45	60
New York	48	62	48	60	..	..	32	56	58	44	60
New Jersey	48	..	50	64	..	..	30	56	56	..	60
New Hampshire	..	60	..	..	..	..	30	56	56	..	60
North Carolina	48	..	50	64	..	..	30	56	54	..	60
North Dakota	48	..	42	60	56	..	32	56	56	..	60
Ohio	48	60	50	60	..	..	32	50	56	45	60
Oklahoma	48	..	42	60	56	..	32	56	56	..	60
Oregon	46	..	42	60	..	..	36	56	56	..	60
Pennsylvania	47	..	48	62	..	..	30	56	56	..	60
South Dakota	48	..	52	60	56	50	32	56	56	..	60
South Carolina	48	60	56	60	..	..	33	56	56	..	60
Vermont	48	64	48	..	60	..	32	56	56	42	60
Virginia	48	60	48	64	..	..	32	56	56	45	60
West Virginia	48	60	52	60	..	..	32	56	56	45	60
Wisconsin	48	..	48	60	..	..	32	56	56	..	60

CHAPTER XVI.

DIFFERENT MAKES OF TRACTION ENGINES.

J. I. CASE TRACTION ENGINES.

These engines are among the simplest and at the same time most substantial and durable traction engines on the market. They are built of the best materials throughout, and are one of the easiest engines for a novice to run.

They are of the side crank type, with spring mounting. The engine is supported by a bracket bolted to the side of the boiler, and a pillow block bearing at the firebox end bolted to the side plate of the boiler.

The valve is the improved Woolf, a single simple valve being used, worked by a single eccentric. The eccentric strap has an extended arm pivoted in a wooden block sliding in a guide. The direction of this guide can be so changed by the reverse lever as to vary the cut-off and easily reverse the engine when desired.

The engine is built either with a simple cylinder or with a tandem compound cylinder.

In the operation of the differential gear, the power is first transmitted to spur gear, containing cushion springs, from thence by the springs to a center ring and four bevel pinions which bear equally upon both bevel gears. The whole differential consequently will move together as but one wheel when engine is moving straight forward or backward; but when turning a corner the four pinions revolve in the bevel gears just in proportion to the sharpness of the curve.

There is a friction clutch working on the inside of the flywheel by means of two friction shoes that can be adjusted as they wear.

There is a feed water heater with three tubes in a watertight cylinder into which the exhaust steam is admitted. The three tubes have smaller pipes inside so that

the feed water in passing through forms a thin cylindrical ring.

The traction wheels are driven from the rims. The
front wheels have a square band on the center of the

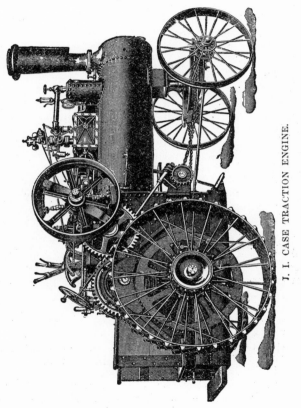

I. I. CASE TRACTION ENGINE.

rim, to prevent slipping sidewise. The smokestack is
cast iron in one piece.

The firebox will burn wood, coal or straw, a fire brick
arch being used for straw, making this fuel give a uniform heat.

The boiler is of the simple locomotive type, with water leg around the firebox and numerous fire flues connecting the firebox with the smokestack in front.   There is safety plug in crown sheet and the usual fittings.   The water tank is under the platform.   The steering wheel and band wheel are on right side of engine.   An independent Marsh pump and injector are used.   The Marsh pump is arranged to heat the feed water when exhaust heater cannot be used.   The governor is the Waters, the safety valve the Kunkle.

THE FRICK CO.'S TRACTION ENGINE.

The most noticeable feature of this engine is that it has a frame mounted on the traction wheels entirely independent of the boiler, thus relieving the boiler of all strain.   This is an undeniable advantage, since usually the strain on the boiler is great enough w i t h o u t forcing the boiler to carry t h e engine and gears.

The gearing to the traction wheels is simple and direct, and a patent elastic

THE FRICK CO.'S TRACTION ENGINE.

spring or cushion connection is used which avoids sudden strain and possible breakage of gears.   Steel traction wheels and riveted spokes.   Differential gear in main axle, with locking device when both traction wheels are required to pull out of a hole.   The reverse gear is single eccentric, the eccentric turning on the shaft.   It is well adapted to using steam expansively.   The crown sheet is so arranged as not to be left bare of water in going up or down hills.   Working parts are covered dust proof.   Engine has self-oiling features and sight feed lubricator.   Friction clutch in flywheel.   Safety brake on main axle.   Engineer's platform mounted on springs and every part of engine requiring attention can be reached conveniently from platform.

Crank is center type. Cross-head pump is used. Usual fittings.

GAAR, SCOTT & CO.'S TRACTION ENGINE.

These engines are built with boiler of locomotive type for burning wood and coal, and of return flue type for burning straw. They are also built of three general types, "Corliss-pattern" frame, "Standard" and "Compound."

The engine is side crank, mounted on brackets attached to the sides of the boiler. The bedplate, cylinder and guides are bored at one operation and cannot get out of alignment. Cylinder has wide ports and free exhaust, and piston has self-setting rings. The genuine link reverse gear is used, as on locomotives, and it undoubtedly has many advantages over any other, including an easily adjustable variable cut-off by correct setting of reverse lever.

The differential gear is heavy and effective. A patent steering attachment, with spiral roll, holds chains taut and gives positive motion. Friction clutch is mounted on engine shaft and connects with the hub of the pinion on this shaft. Rigid pinion is also provided. Cross-head pump and injector are used, and Pickering governor with improved spring speeder, permitting quick and easy change of speed; also Sawyer's lever for testing safety. Steam passes direct from dome to cylinder, without loss from cooling or condensing. The steel water tank can be filled by a jet pump operated by steam.

D. JUNE & CO.'S TRACTION ENGINE.

This is one of the very few traction engines built with upright boiler, but it has been on the market many years and has been widely used with great success as a general road locomotive.

The engine is mounted on the water tank. The weight of the boiler comes on the hind wheels, and makes this type of engine superior for pulling. It is claimed that it has no equal on the market as a puller. The upright type of boiler has the advantage that the crown sheet is never exposed and it is claimed flues will last longer than in horizontal type. It works equally well whether it stands level or not, an advantage that no other type has.

This type gets up steam more quickly than any other—it is said, from cold water, in twenty minutes. The steam is superheated in a way to economize fuel and water.

By being mounted on the tank, the engine does not get hot as it would if mounted on the boiler, and the corresponding straining of parts is avoided. A patent water spark arrester is used which is an absolute protection.

D. JUNE & CO.'S TRACTION ENGINE.

The engine is geared to the traction by a chain, which can easily be repaired as the links wear. The friction clutch works inside flywheel. Engine has a new reversible eccentric, and differential gear, with usual fittings.

NICHOLS & SHEPARD TRACTION ENGINE.

The builders of this engine lay special stress upon the care with which the boiler and similar parts are constructed. The important seams are double riveted, and

NICHOLS & SHEPARD TRACTION ENGINE.

the flue sheet is half inch steel, drilled instead of punched for the flues, and fitted with seamless steel flues, all of the best steel.

The boiler is the direct flue locomotive type. The crown

sheet slopes backward to allow it to be covered with water in descending hills. Boiler has round-bottom firebox. Axle passes around below the boiler, and springs are provided.

The engine is mounted on a long heater, which is attached to the side of the boiler. The locomotive link reverse is used, with a plain slide valve.

Cross-head pump and injector are used, and improved pop safety valve. Cylinder is jacketed, and cross-head guides are rigid with cylinder, so that perfect alignment is always secured.

Engines are built to burn coal or wood. A straw burner is provided with firebrick arch. Compound engines are also built.

THE HUBER TRACTION ENGINE.

The Huber boiler is of the return flue type, and the gates are in the large central tube. This does away with the low-hanging firebox, and enables the engine to cross streams and straddle stumps as the low firebox type cannot do. The cylindrical shape of the boiler also adds considerably to its strength. The water tank is carried in

THE HUBER TRACTION ENGINE.

front, and swings around so as to open the smoke box, so that repairs may be made on the fire tubes at this end easily in the open air. With water front return flue boilers the workman has to crawl through entire length of central flue. As there is no firebox, the boiler is mounted above the axle, not by bolting a plate to the side of the firebox. The boiler is made fast to the axle, which is mounted on wheels with spring cushion gear, the springs being placed in the wheel itself, between the two bearings of the wheel or the hub on trunnions, which form the spindle for the hub. The wheel revolves on the trunnion instead of on the axle, and there is no wear on the axle. The traction gear has a spring connection so that in starting a load there is little danger of breakage. The compensating gear is all spur. The intermediate gear has a ten-inch bearing, with an eccentric in the center for adjusting the gear above and below. There is a spring draw bar and elastic steering device. An improved friction clutch works on inside of flywheel. Engine has a special governor adapted to varying work over rough roads, etc.

A single eccentric reverse gear is used, with arm and wood slide block (Woolf); and there is a variable exhaust, by which a strong draft may be quickly created by shutting off one of two exhaust nozzles. When both exhausts are open, back pressure is almost entirely relieved.

The steam is carried in a pipe down through the middle of the central flue, so that superheating is secured, which it is claimed makes a saving of over 8 per cent in fuel and water. The stack is double walled with air space between the walls.

A special straw-burning engine is constructed with a firebox extension in front, and straw passes over the end of a grate in such a way as to get perfect combustion. This make of engine is peculiarly adapted to burning straw successfully.

A. W. STEVENS' TRACTION ENGINE.

This engine has locomotive pattern boiler, with sloping crown sheet, and especially high offset over firebox, doubling steam space that will give dry steam at all times. A large size steam pipe passes from dome in rear through boiler to engine in front, superheating steam and avoiding condensation from exposure. Grate is a rocking one, easily cleaned and requiring little attention, and firedoor is of a pattern that remains air-tight and need seldom be opened.

The engine is mounted upon the boiler, arranged for rear gear traction attachment. Engine frame, cylinder, guides, etc., are cast in one solid piece.

It has a special patented single eccentric reverse, and

A. W. STEVENS' TRACTION ENGINE.

Pickering horizontal governor. There is a friction clutch, Marsh steam pump, and injector. Other fittings are complete, and engine is well made throughout.

AULTMAN-TAYLOR TRACTION ENGINE.

The Aultman-Taylor Traction Engine is an exceptionally well made engine of the simplest type, and has been on the market over 25 years. There are two general types, the wood and coal burners with locomotive boilers, and return flue boiler style for burning straw. A compound engine is also made with the Woolf single valve gear.

A special feature of this engine is that the rear axle comes behind the firebox instead of between the firebox and the front wheels. This distributes the weight of the engine more evenly. The makers do not believe in springs for the rear axle, since they have a tendency to wear the gear convex or round, and really accomplish much less than they are supposed to.

Another special point is the bevel traction gear. The engine is mounted on the boiler well toward the front,

AULTMAN-TAYLOR TRACTION ENGINE.

and the flywheel is near the stack (in the locomotive type). By bevel gears and a long shaft the power is conducted to the differential gear in connection with the rear wheels. The makers claim that lost motion can be taken up in a bevel gear much better than in a spur gear. Besides, the spur gear is noisy and not nearly so durable. Much less friction is claimed for this type of gear.

The governor is the Pickering; cross-head pump is used, with U. S. injector, heater, and other fittings complete. A band friction clutch is used, said to be very durable. Diamond special spark arrester is used except in straw burners. The platform and front bolster are provided with springs. The makers especially recommend their compound engine, claiming a gain of about 25 per cent. The use of automatic band cutters and feeders, automatic weighers and baggers, and pneumatic stackers with threshing machine outfits make additional demands on an engine that is best met by the compound type. With large outfits, making large demands, the compound engine gives the required power without undue weight.

AVERY TRACTION ENGINE.

The Avery is an engine with a return flue boiler and full water front, and also is arranged with a firebox besides. There is no doubt that it effects the greatest economy of fuel possible, and is adaptable equally for wood, coal, or straw. The boiler is so built that a man may

AVERY TRACTION ENGINE.

readily crawl through the large central flue and get at the front ends of the return tubes to repair them.

The side gear is used with a crank disc instead of arm. The reverse is the Grime, a single eccentric with device for shifting for reverse. The friction clutch has unusually long shoes, working inside the flywheel, with ample clearance when lever is off. A specialty is made of extra wide traction wheels for soft country. The traction gear is of the spur variety. There is also a double speed device offered as an extra.

The water tank is carried in front, and lubricator, steering wheel (on same side as band wheel for convenience in lining up with separator), reverse lever, friction clutch, etc., are all right at the hand of the engineer.

The traction gear is of the spur variety, adjusted to be evenly distributed to both traction wheels through the compensating gear, and to get the best possible pull in case of need.

For pulling qualities and economy of fuel, this engine is especially recommended.

BUFFALO PITTS TRACTION ENGINE.

The Buffalo Pitts Engine is built either single cylinder or double cylinder. The boiler is of the direct flue locomotive type, with full water bottom firebox. The straw burners are provided with a firebrick arch in the firebox. Boilers are fully jacketed.

BUFFALO PITTS TRACTION ENGINE.

The single and double cylinder engines differ only in this one particular, the double cylinder having the advantage of never being on a dead center and starting with perfect smoothness and gently, seldom throwing off belt. The frame has bored guides, in same piece with cylinder, effecting perfect alignment.

The compensating gear is of the bevel type, half shrouded and so close together that sand and grit are kept out. Three pinions are used, which it is claimed prevent rocking caused by two or four pinions.

Cross-head has shoes unusually long and wide. The engine frame is of the box pattern, and is also used as a heater, feed water for either injector or steam pump passing through it. Valve is of the plain locomotive slide type.

The friction clutch has hinged arms working into flywheel with but slight beveling on flywheel inner surface, and being susceptible of easy release. It is a specially patented device. The Woolf single eccentric reverse gear is used. Engine is fully provided with all modern fittings and appliances in addition to those mentioned. It was the only traction engine exhibited at Pan-American Exposition which won gold medal or highest award. It claims extra high grade of workmanship and durability.

THE REEVES TRACTION ENGINES.

These engines are made in two styles, simple double cylinder and cross compound. The double cylinder and cross compound style have been very successfully adapted to traction engine purposes with certain advantages that no other style of traction engine has. With two cylinders and two pistons placed side by side, with crank pins at right angles on the shaft, there can be no dead centers, at which an engine will be completely stuck. Then sudden starting is liable to throw off the main belt. With a double cylinder engine the starting is always gradual and easy, and never fails.

The same is equally true of the cross compound, which has the advantage of using the steam expansively in the low pressure cylinder. In case of need the live steam may b⟂ introduced into the low pressure cylinder, enormously increasing the pulling power of the engine for an emergency, though the capacity of the boiler does not permit long use of both cylinders in this way.

THE REEVES TRACTION ENGINE.

The engine is placed on top of the firebox portion ot the boiler, and the weight is nicely balanced so that it comes on both sides alike.

The gearing is attached to the axle and countershaft which extend across the engine. The compensating gear is strong and well covered from dirt. The gearing is the gear type, axle turning with the drivers. There is an independent pump; also injector, and all attachments. The band wheel being on the steering wheel or right side of the engine, makes it easy to line up to a threshing machine. Engine frame is of the Corliss pattern; boiler of locomotive type, and extra strongly built.

THE RUMELY TRACTION ENGINE.

The most striking peculiarity is that the engine is mounted on the boiler differently from most side crank traction engines, the cylinder being forward and the shaft at the rear. This brings the gearing nearer the traction wheels and reduces its weight and complication.

THE RUMELY TRACTION ENGINE.

The boiler is of the round bottom firebox type, with dome in front and an ash pan in lower part of firebox, and is unusually well built and firmly riveted.

The traction wheels are usually high, and the flywheel is between one wheel and the boiler.

The engine frame is of the girder pattern, with overhanging cylinder attached to one end.

The boiler is of the direct flue locomotive type, fitted for straw, wood, or coal. Beam axle of the engine is behind the firebox, and is a single solid steel shaft. Front axle is elliptical, and so stronger than any other type.

A double cylinder engine is now being built as well as the single cylinder. The governor regulates the double cylinder engine more closely than single cylinder types, and in the Rumely is very close to the cut-off where a special simple reverse is used with the double cylinder engine.

Engine is supplied with cross-head pump and injector, Arnold shifting eccentric reverse gear, friction clutch, and large cylindrical water tank on the side. It also has the usual engine and boiler fittings.

PORT HURON TRACTION ENGINE.

The Port Huron traction engine is of the direct flue locomotive type, built either simple or compound, and of medium weight and excellent proportions for general purpose use. The compound engine (tandem Woolf cylinders) is especially recommended and pushed as more economical than the simple cylinder engine. As live steam can be admitted to the low pressure cylinder, so turning the compound into a simple cylinder engine with two cylinders, enormous power can be obtained at a moment's notice to help out at a difficult point.

Two injectors are furnished with this engine, and the use of the injector is recommended, contrary to the general belief that a pump is more economical. The company contends that the long exhaust pipe causes more back pressure on the cylinder than would be represented by the saving of heat in the heater. However, a cross-head pump and special condensing heater will be furnished if desired.

On the simple engine a piston valve is used, the seat of the valve completely surrounding it and the ports being circular openings, the result, it is claimed, being a balanced valve.

The valve reverse gear is of the Woolf pattern, the engine frame of the girder type, Waters governor, with special patent speed changer, specially balanced crank disc, patent straw burner arrangement for straw burning engines, special patent spark extinguisher, special patent gear lock, and special patents on front axle, drive wheel and loco cab.

The usual fittings are supplied.

PORT HURON TRACTION ENGINE.

MINNEAPOLIS TRACTION ENGINE.

The Minneapolis traction engine is built both simple

MINNEAPOLIS TRACTION ENGINE.

and compound. All sizes and styles have the return flue boiler, for wood, coal or straw. Both axles extend entirely and straight under the boiler, giving complete support without strain. The cylinder, steam chest and guides form one piece, and are mounted above a heater, secured firmly to the boiler; valve single simple D pattern. Special throttle of the butterfly pattern, large crank pin turned by special device after it is driven in, so insuring perfect adjustment; special patent exhaust nozzle made adjustable and so as always to throw steam in center of stack; friction clutch with three adjustable shoes. Boiler is supplied with a superheater pipe. Woolf valve and reverse gear. Special heavy brass boxes and stuffing-boxes. Sight feed lubricator and needle feed oiler; Gardner spring governor. Complete with usual fittings. This is a simply constructed but very well made engine.

INDEX

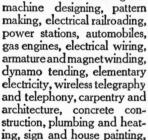

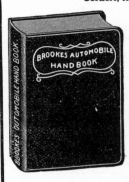

The Practical Gas & Oil Engine HAND-BOOK

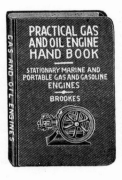

A MANUAL of useful information on the care, maintenance and repair of Gas and Oil Engines.

This work gives full and clear instructions on all points relating to the care, maintenance and repair of Stationary, Portable and Marine, Gas and Oil Engines, including How to Start, How to Stop, How to Adjust, How to Repair, How to Test.

Pocket size, 4 x 6½.

232 pages. With numerous rules and formulas and diagrams, and over 70 illustrations by L. ELLIOTT BROOKES, author of the "Construction of a Gasoline Motor," and the "Automobile Hand-Book."

This book has been written with the intention of furnishing practical information regarding gas, gasoline and kerosene engines, for the use of owners, operators and others who may be interested in their construction, operation and management.

In treating the various subjects it has been the endeavor to avoid all technical matter as far as possible, and to present the information given in a clear and practical manner.

16mo. **Popular Edition—Cloth. Price**..................$1.00
Edition de Luxe—Full Leather Limp. Price......... 1.50

FREDERICK J. DRAKE & CO.
PUBLISHERS